Radiation: An Energy Carrier

Tetsuo Tanabe

Radiation: An Energy Carrier

 Springer

Tetsuo Tanabe (iD)
Research Center for Artificial
Photosynthesis
Osaka Metropolitan University
Osaka, Japan

Emeritus Professor
Kyushu University
Fukuoka, Japan

Emeritus Professor
Nagoya University
Nagoya, Japan

ISBN 978-981-19-1956-5   ISBN 978-981-19-1957-2   (eBook)
https://doi.org/10.1007/978-981-19-1957-2

Translation from the Japanese language edition: *Enerugi no Shiten kara Mita Houshasen* by Tetsuo Tanabe, © Kyushu University Press 2018. Published by Kyushu University Press. All Rights Reserved.
© Kyushu University Press 2022

This Springer imprint is published by the registered company Springer Nature Singapore Pte Ltd.
The registered company address is: 152 Beach Road, #21-01/04 Gateway East, Singapore 189721, Singapore

Preface to English Edition

After nuclear accidents such as Three Mile Island in USA, Chernobyl in Ukraine, and Fukushima Nuclear Power Plant in Japan, many people were evacuated and lost their homes, because of radioactive fall-out from the damaged reactors. Even today, a number of people cannot return to the places they lived before the accident. Although the initial impact of the accidents was quite large, the harm caused by exposure to radiation seems gradually disappearing. Nevertheless, these accidents make people more nervous of the exposure to radiation. The number of people who feel that radiation is "scary/fearful" has increased, along with the number of people who refuse to accept nuclear power plants. In fact, all of the nuclear power plants in Japan were stopped for several years after the Fukushima accident at 2011. At that time, a shortage of electricity could be avoided, because the electricity generated by the nuclear plants has been replaced by that generated in thermal power plants burning coals and oil. The coal and oil plants are not desirable due to environmental impacts and resource issues, but people accepted this provisional measure.

In recent years, due to concerns about global warming, many governments have decided not to build new power plants burning coal and tried to replace them with renewable energy mostly solar cells and window power. Any power source has disadvantages or creates some risks. The number of people who is affected by global warming is far larger than the number of people who suffered in nuclear accidents. Generally, the negative effects or demerits of using powerful technology necessary to sustain modern civilization seem to be compensated by insurance. The typical example is traffic accidents; the number of people involved is quite large. Nevertheless, most people accept the risk of traffic accident and compensate their potential harm by money from insurance. On the other hand, the risk to use nuclear power seems to be discussed on completely different base from other risks accompanying with utilization of energy or power. Many people refuse to accept nuclear reactors. Probably that is because they are afraid of after-effects of exposure to radiation which could appear for themselves later, on their children through influence of genes, cancer risk in offspring and so on, which most people believe, cannot be covered by insurance.

Utilization of large energy or power always accompanies some risk when the energy or power is released by accidents in a different way from the original purpose. Therefore, the risk should be shared by the beneficiaries. One of the main reasons for hesitating to use nuclear power is that people have not accepted this principle, or the risk of radiation is considered to be an exception from the principle. There seem two major reasons for that. One is that people fair "radiation" without having clear reason. The other is that "radiation" is mysterious and it is difficult to evaluate the risk, i.e., what kind of effects occur and how the effects of the radiation exposure appear.

Therefore, it is quite natural that there are pros and cons on the acceptance of nuclear energy. Unfortunately, the debate is often conducted without sufficient knowledge or correct understanding of radiation. The debate should consider long-term energy security with sufficient knowledge or correct understanding of radiation.

Based on above-mentioned general background, the first Japanese edition of this book was published aiming at assisting readers (1) to understand "radiation," i.e., what radiation is or means and why radiation is dangerous, (2) to get correct understanding that the radiation is carrying energy, and (3) to accept that the radiation is not "scary" but "dangerous but controllable and useful." Hopefully, this English translation will help readers to do so. Chinese translation will be also published by University of Science and Technology of China Publishing House as ISBN 978-7-312-05302-3.

The author very much appreciates Dr. Richard More for his review of English writing.

Osaka, Japan   Tetsuo Tanabe
February 2022

Preface

The purpose of this book is to explain "radiation" from somewhat different viewpoint than the traditional images like "radiation is incomprehensible, scary, fearful, dangerous, hazardous, etc.," and to give a more accurate understanding that the radiation carries energy, and to help people realize that the radiation is not "scary" but rather "dangerous but well-understood, controllable, and can be useful to humanity."

Many introductory books or textbooks have been published on subjects such as radiation physics, radiochemistry, radiobiology, and radiolysis. In recent days, information about radiation can be obtained very easily on various websites. Some of these books and websites are listed in bibliography at the end to this book. The author has obtained useful information from these sources. Nevertheless, there are not many books that discuss the effect of exposure to radiation, based on the idea that radiation carries energy and the exposure to radiation means that energy is deposited or absorbed in an exposed object. In many published books, the biological effects of the exposure to radiation are the main themes using a specified term of "absorbed dose," "dose equivalent," or "equivalent dose," which often enhances the feeling that "radiation is scary." Furthermore, the effects caused by a lower dose of the radiation exposure that are difficult to see are often not discussed.

The aim of this book is to help readers to understand "radiation" from a broader viewpoint, focusing on the fact that radiation carries energy, and deposits energy in an exposed object. What effects appear after the exposure and why these effects appear are quite differently depending on characters or properties of the object.

Although the contents may be not easy to understand without knowledges of mathematics, chemistry, and physics, as taught in high school, the author tries to explain or describe radiation science as simply as possible and to give correct information for understanding radiation.

Chapter 1 describes what is written in the book, in particular, what "radiation" is or means. The following chapters provide detailed descriptions necessary to correctly understand radiation. All chapters are to be independent for easy reading so that readers can start with any chapter among Chaps. 2–9 with free choice. It is not necessary to read them sequentially.

Chapter 2 explaines that radiation is consisting of high energy particles and/or electromagnetic waves referred as energetic quanta (EQ) or quantum particles. Accordingly, in the most of the book, "radiation" is referred as EQ. In Chap. 3, details of EQ sources are described. Chap. 4 shows the effects of EQ exposure mostly appears as damage in objects, and this is discussed separately for inorganic and organic materials, and for living beings, based on physical and chemical processes of damaging and recovering. Chapter 5 describes reduction of absorbed dose including shielding and decontamination. The detection and measurement of EQ, which are essential for the discussion of EQ, are summarized in Chap. 6. Chapter 7 provides examples of how Q are used in various fields in view of the theme of this book, "radiation carries energy." In Japan, the use of EQ in the medical field is progressing, and the average absorbed dose of EQ for medical purposes has reached more than half of the annual EQ exposure in nature. Details of the medical use of EQ can be found in many books published (some are listed at the end).

Once the readers have understood that radiation consists energetic quanta (EQ) and carries energy, they note an interesting connection with the history of life on the earth and energy storage and consumption, which is described in Chap. 8. The development of living beings in the earth's history has been influenced by EQ from the sun. Chapter 9 describes "energy and radiation" focusing on the energy sources and shows that risks accompany the utilization of all energy sources.

It should be noted that the use of solar energy is a use of "radiation," and that the sun itself and the Earth's atmosphere are converting high-energy radiation, which is dangerous to mankind, into low-energy and useful radiation.

The author also hopes that after reading this book, readers can agree that although radiation (EQ) is dangerous or sometimes toxic, it is controllable and should be used as a long-term energy source which will allow human beings continue to exist on the earth. Since utilization of nuclear energy as an energy source for the earth is realized in nature as solar energy, it is logical that nuclear power, either fission or fusion reactors, could be artificial energy sources if there is sufficient effort to ensure their safety. Nevertheless, it should be realized that when using a large amount of energy, some risk is always present. The risks have appeared as pollution and global warming caused by waste heat and exhaust gases from the power plants burning oils or coals. There is no doubt that nuclear energy can be used as an energy source in long term, but it is mandatory to consider how to handle the risks that go with it.

Osaka, Japan   Tetsuo Tanabe
December 2017

Contents

Chapter 1
Radiation Carries Energy

Abstract The main subject of this book is to introduce radiation as an energy carrier and being dangerous but controllable and useful. The purpose is to remove scary feelings on radiation, with making people understand what radiation is or means and why it is dangerous or hazardous. An additional target is to convince the readers that it is possible to avoid the radiation exposure and the radiation can be controllable and used beneficially as energy sources under safety regulations. The first chapter is the general introduction on what is written in this book.

Keywords Dose · Dose equivalent · Energetic quanta · Energy carrier · Exposure · Radiation

1.1 Is "Radiation" Scary?

Most people say that "radiation" is scary, or they fear it. However, if they were asked "why", little could answer and most would say "it is based on common knowledge", or "it is dangerous but we do not know why or what is the radiation". Or intellectual people may say "Because it is difficult to estimate what kind of and how effects appear on the radiation exposure to human beings".

There is no wonder that people fear the radiation when they see the appearance of damage on residents suffered by nuclear bombs in Nagasaki and Hiroshima, Japan, and world nuclear bomb tests in open air. They were exposed to the radiation with doses ranging from very low levels to over lethal doses. However, effects of the radiation exposure with the dose level of lower than 1 mSv hardly appear or are very much scattered from mostly nothing to a few in canceration for example.

After the accident at Fukushima Daiichi Nuclear Power Plant (Fukushima Nuclear Power Plant, Tokyo Electric Power Company), many residents were evacuated and lost their place to live. This makes people more nervous on the radiation exposure. Still, not a small number of the suffered residents could not come back to their places. Although the initial impact of the accidents was quite large, the direct impact of the radiation exposure seems gradually disappear. And the number of people directly influenced by the radiation exposure caused by radioactive materials released by the accident was not so large. Nevertheless, the accident makes people more nervous

T. Tanabe, *Radiation: An Energy Carrier*,
https://doi.org/10.1007/978-981-19-1957-2_1

of the radiation exposure and the number of people who decline to accept nuclear power plants has increased.

After the accident, all nuclear power plants in Japan were stopped. Nevertheless, the shortage of electricity could be avoided because most of the electricity generated by the nuclear plants has been replaced by that generated by thermal power plants burning coals or oils which are not desirable due to environmental and resource issues but people have accepted.

Recent global warming is attributed to carbon dioxide (CO_2), though some scientists believe that it is not due to CO_2. The global warming is influencing the earth and will be more significant in the future. However, current people's concern on the burning fossil fuels seems much less compared to those on utilization of the nuclear power. Probably because any countermeasure to mitigate the global warming is believed to be taken. Does this mean that concerns on the global warming are less than those on utilization of nuclear power? Anyway, the purpose of this book is not to discuss what should be the energy sources, which are revisited in Chap. 9.

At the Fukushima accident, people lived within about 20 km from the Fukushima power plants or areas showing high radiation levels (air dose) were evacuated. The high air radiation dose area is distinguished from other areas based on absorbed dose rate or dose equivalent rate. In scientific fields handling radiation, such as medicine, health physic, radiobiology, radiology, and radiation chemistry, the term "**absorbed dose**" is used instead of "**absorbed energy**" to feature the energy given to living beings or specifically human beings by exposure of radiation using the unit of Gray (Gy). The absorbed dose equivalent given with the unit of Sievert (Sv) was introduced to normalize different effects of the kinds of radiation. These units are described in Sect. 1.2.4 in detail.

To avoid the radiation exposure is to keep away from radiation sources or materials including radioisotopes (RIs), or to install a shield that does not allow the radiation to pass through. Although the evacuation of residents near the Fukushima power plants was an unavoidable procedure, it was difficult to determine a critical absorbed dose rate above which the evacuation was required. (There is no so-called threshold in the absorbed dose rate to distinguish whether it is dangerous or not.) People tend to set the critical level to give less influence, even if it is too low. Ironically if the level was set lower, more people would fear the radiation exposure.

Often appeared is a rumor or misunderstanding that when a substance was exposed to radiation, it has been changed to be radioactive. This never happens and is completely wrong. The radioactivity is never transferred to the substance exposed to the radiation, totally different from the transfer of virus, the cause of disease. However, the rumor that people exposed to the radiation became radioactive unintentionally spread and those who believed the rumor being true often discriminates people evacuating from Fukushima. That is a very unfortunate event and should be avoided. Nevertheless, it is quite hard to delete the rumor once circulated and to convince people not to believe the rumor. One of the purposes of this book is to avoid such misunderstandings on the radiation.

There is no doubt that it is important to understand the radiation and correct understanding makes it possible to manage or handle the radiation safely. Nevertheless, it is not easy to understand the radiation correctly or even to know whether one correctly understands it.

As for "radiation", many books have been published and some are listed at the end of this book as bibliography. Majority of the published books are on radiobiology and radiology discussing the biological effects of radiation exposure on living beings. The observable biological effects (damages and diseases) caused by the radiation exposure and the recovery of the damages and diseases are main targets. In most cases, the damages and diseases clearly observed are discussed. However, the effects caused by the radiation exposure starts in very tiny area with nm scales, and are hence invisible, until the damage area appears in cells and tissues. The processes to become visible damages are not explained well except the damaging processes in cells, in particular, DNA and genes. Therefore, the more one studied radiobiology, stronger impressions he would have that the radiation is scary. Once one feels scared on the radiation without any evidence, it is not easy to change his feeling by others. To avoid wrong rumors or misunderstandings on the effects of radiation exposure it is critically important to understand the radiation correctly, and the correct understanding makes possible to manage or handle the radiation safely. Nevertheless, it is not easy to understand the radiation correctly or even to know whether one correctly understands it.

Based on the above-mentioned background, the purpose of this book is to remove scary feelings on radiation, with making people understand what radiation is or means and why it is dangerous or hazardous. An additional target is to convince the readers that it is possible to avoid the radiation exposure and the radiation can be controllable and used beneficially as energy sources under safety regulations.

1.2 What is Written in This Book?

This book describes what is radiation or what radiation means, and the effects of radiation exposure on substances. The word "radiation" is generically used to represent high-energy particles like electrons and ions and electromagnetic waves. Exposure (to radiation) means that some or all of energy carried by the radiation is deposited to (absorbed in) a substance exposed to the radiation. The manner of the energy deposition or absorption in the substance by the exposure is quite different depending on types of the radiation, their carrying energy, and characters or properties of the substance. The term "exposure" is often limitedly used for the energy deposition or absorption in a human or human body exposed to the radiation. The effect of the heavy exposure of a human being appears as some disease like cancer, or even death of tissues, organs, and a human body, which makes people scared of the radiation.

Above 0 K, any matter including living beings releases and absorbs energy from its surrounding. The temperature of the matter is decided by the balance of the release and absorption of energy. To be exact, the temperature is determined by the power balance of input and output. Energy and power are often used in confusion and sometimes misused in discussion of radiation effects. However, both are different in definition. The power is the amount of energy transferred or converted per unit of time and is transferred with the form of radiation (of electromagnetic waves), and conduction and convection of heat. The first one is well known as blackbody radiation which occurs even in vacuum, while the latter two require surrounding materials. Examples of the blackbody radiation are seen in infrared heaters or warming, and a thermometer to measure the temperature of a human body which is routinely used in airports. Heat conduction relies on energy exchange between mutually contacting materials through atomic and molecular motions. Convection is heat transport accompanied with thermal motion of fluid.

Radiation, as described later, carries energy as electromagnetic waves or particles, and transfers or interchanges its energy when it collides with or enter a substance. The energy transferred in unit time is the power. It may sound strange to hear that "energy transfer/interchange with radiation". This means that if high-energy radiation enters a substance, energy is given to the substance or the substance absorbs energy, while if low energy radiation enters a substance, the radiation is given energy from the substance, which is just the physical phenomena of heating and cooling of the substance by the radiation. Thus "radiation" discussed in this book is high-energy electromagnetic waves or particles. And deposition or absorption of all or some of their energy in human body under the exposure is "radiation exposure" that people fear.

The manner or process of energy deposition or absorption depends entirely on the types of radiation, its energy and intensity, and the nature of the substance, particularly its density and temperature. Furthermore, the energy deposition results in damage to the substance, i.e., the formation of defects, or disorders. While the substances whatever they are, living beings, organic materials, and inorganic materials have resilience to recover the damages, which makes the appearance of the damages or the effect of exposure complex as discussed in Chap. 4.

The energy range of radiation spreads very wide-ranging from 0 eV to more than 10^{20} eV as shown in Table 1.1 for the electromagnetic wave as an example of the radiation. Dangerous or scary radiation is those having the energy of above a few eV, while those having less are generally not hazardous.

The following sections in this chapter are devoted to explaining following nine subjects. In Sect. 1.2, four characteristics of radiation are introduced

1. Radiation is carrying energy and transfers energy as power,
2. Energy and intensity of radiation,
3. Radiation effect is different depending on the kind of radiation even if they carry the same energy, and
4. Physical units relating to radiation and radiation measurements.

In the remaining sections are summarized five important points of this book as an introduction to the radiation.

5. Energy emission from materials (black body radiation and radiation from radioactive materials),
6. The universe and natural radiation,

Table 1.1 Energy carried by EQ (electromagnetic waves) and corresponding wavelength and frequency

Energy		Frequency (Hz)	Wave length	Name	Purpose in use
10 MeV	Ionizing radiation			γ-ray	Medical treatment
100 keV		3 EHz		X-ray	Non-destructive inspection X-ray photography
1 keV		300 PHz	1 nm	Ultraviolet light	Disinfection
10 eV		3 PHz	100 nm		
				Visible light	
0.1 eV	Non-ionizing radiation	30 THz	10 μm		
10 meV		3 THz	100 μm	Infrared light	Heater
1 mcV		300 GHz	1 mm	Sub-millimetric wave	
0.1 meV		30 GHz	1 cm	Millimetric wave	Radar
10 meV		3 GHz	10 cm	Centimetric wave	Satellite communication
1 meV		300 MHz	1 m	Ultra-high frequency wave (UHF)	Microwave
0.1 meV		30 MHz	10 m	Very high frequency wave (VHF)	FM broadcast, TV
10 neV		3 MHz	100 m	Short wave	Radio communication, Shortwave broadcast
1 neV		300 kHz	1 km	Medium frequency wave	AM broadcast, Ham radio
0.1 neV		30 kHz	10 km	Ultra-short wave	Marine radiocommunication
10 peV	Electromagnetic waves	3 kHz	100 km	Very low frequency wave	Long distance communication
0.1 peV		60/50 Hz	10 Mm	Commercial electricity	

1 EHz $= 10^{18}$ Hz, 1 PHz $= 10^{15}$ Hz, 1 THz $= 10^{12}$ Hz, 1 GHz $= 10^{9}$ Hz, 1 MHz $= 10^{6}$ Hz, 1 kHz $= 10^{3}$ Hz, 1 neV $= 10^{-9}$ eV, 1 peV $= 10^{-12}$ eV

7. Energy of radiation,
8. Radioactive materials in nature and artificial radiation sources,
9. Radiation shielding.

1.2.1   Radiation is Carrying Energy

Radiation is a general phenomenon indicating emission of light or particles from materials in higher energy states. However, in radiation safety, the radiation is limitedly used to represent those that could give some influence on living beings when they are exposed to it. Such radiation consists of energetic particles and/or electromagnetic waves (light or photons). The behavior of the energetic particles and photons is controlled by quantum physics, and they are unified as energetic quanta. Hence it would be better to say that "radiation" is the energetic quanta. In most parts of this book, the term of **(EQ) Energetic Quanta** is used instead of "radiation" to make the meaning of the radiation definite.

There are two ways to carry or transfer energy in nature: one is as the kinetic energy of particles (atoms, nuclei, electrons, etc.), and the other as electromagnetic waves (photons). Figure 1.1 shows how energy is carried by a particle and a photon. For any moving particle (from very tiny one like an electron and atom to quite large one like the earth and the sun) with a velocity, v_i, and mass, m_i, its carrying energy, ε_i, is given as the kinetic energy of its translational motion,

$$\varepsilon_i = \frac{1}{2}m_i v_i^2 \tag{1.1}$$

In a material consisting of various atoms, they carry additional energy as internal energy such as rotation and vibration of the atoms, which will be discussed in later chapters. Although materials made up of different sizes and weights are named differently, like the earth, a human being, molecules, atoms, elementary particles,

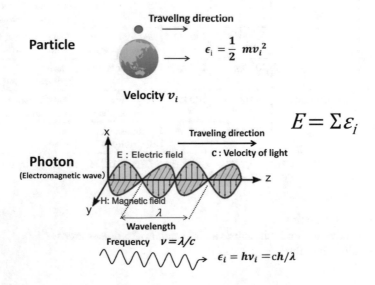

Fig. 1.1 Energy carried by a particle and an electromagnetic wave (photon)

and EQ like α- and β-particles, the energy carried by their translational motion is represented by the same equation as Eq. (1.1).

Different from the particle, energy carried by a photon (an electromagnetic wave) with frequency of ν_i and wavelength of λ_i is given by

$$\varepsilon_i = h\nu_i = \frac{ch}{\lambda_i}, \tag{1.2}$$

where h and c are respectively the Planck constant, $6.62607004 \times 10^{-34}$ J s^{-1} and the velocity of light, 2.9979×10^8 m s^{-1}. Irrespective of the magnitude of the frequency and wavelength, the energy carried by the photon is given Eq. (1.2).

To discuss the magnitude of energy, various energy units are introduced. The most commonly used is calories (cal). In physics and chemistry, joules (J) are used. For discussion of EQ (radiation), electron volts (eVs) are often used. The conversion among cal, J and eV is given by

$$1 \text{ J} = 0.24 \text{ cal} = 6.24 \times 10^{18} \text{ eV}. \tag{1.3}$$

The significantly different digits of eV from those of J and cal are due to the difference in masses handled, i.e., Avogadro number of 6×10^{23} atoms and molecules for one mole is included in molar unit in chemistry (J and cal), whereas eV is for handling one quantum-like an electron, an atom, a molecule and so on. Using the gas constant R, the temperature of a material, T, can be converted to its energy, E, as

$$E = RT, \tag{1.4}$$

where R is the gas constant;

$$R = 8.3 \text{ J K}^{-1} \text{ mol}^{-1} = 1.38 \times 10^{-23} \text{ J K}^{-1} = 8.6 \times 10^{-5} \text{ eV K}^{-1}. \tag{1.5}$$

J K^{-1} mol^{-1} is for a unit mole, and J K^{-1} and eV K^{-1} for individual particles. Then the temperature of 1 mol of a material constructed of atoms or molecules having the energy of 1 eV, is calculated to be around 10^4 K as given by

$$T = 1(\text{eV}) \div 8.6 \times 10^{-5}(\text{eV K}^{-1}) = 1.1 \times 10^4(\text{K}). \tag{1.6}$$

Particles and electromagnetic waves having higher energy are quantized and referred to as quantum particles and photons, respectively, or unified as quanta and that is the reason to use EQ (Energetic Quanta) instead of the radiation in this book.

Electromagnetic waves (photons) are differently named corresponding to their carrying energy or wavelength/frequency as summarized in Table 1.1 and Fig. A.5 in Appendix (Q&A), from higher to lower energy, as γ-photon, X-ray, ultraviolet light, visible light, infrared light, macro-wave, radio wave, and so on. (Although usually "γ-ray" is used, "γ-photon" is used throughout this book in order to make clear that γ-ray is a kind of EQ.) Historically, each of them had been studied independently and understood separately until the twentieth century when quantum mechanics was established and has proved that all are electromagnetic waves with different wavenumbers or frequencies and unified as photons. In Table 1.1, one can see that they are ranging over 10^{15} orders of magnitude.

Quantum mechanics makes it clear that electromagnetic waves behave like particles when their energy is very high, while very high energy particles behave like electromagnetic waves, and they can be converted to each other as quanta. Atoms, electrons, neutrons, and other elementary particles are all quanta with mass. Electromagnetic waves are also quanta but have no mass, and are collectively referred to as photons. "Radiation" includes all EQ as a general term. As described above, when the energy of the electromagnetic wave is very high, it behaves like a particle named photon. The minimum energy carried by one quantum is about the Plank's constant 10^{-34} J (10^{-15} eV), the maximum is infinite. Usually, EQ above keV (10^3 eV) is dangerous for living beings. Even with such large difference in energy over 10^{10} order, the formula giving the EQ energy is the same, either Eq. (1.1) or Eq. (1.2). However, if the energy of quanta is small, it is impossible to identify each quantum. Similar observation often appears, for example, when you are dealing with water, you cannot distinguish each constituent H_2O molecule. Sound also travels as waves and is quantized as a phonon in quantum physics. However, the energy carried by the phonon is so small as less than 10^{-3} eV that the phonon is not dangerous and not handled in this book.

1.2.2 All Physical and Chemical Phenomena Accompany Energy Transfer

Any physical and chemical phenomena always accompany transfer or exchange of energy, and their occurring in space and time is different depending on the amount of energy transferred or exchanged. Table 1.1 shows how photons are used depending on their wavelength/frequency. That is because depending on their energy range, physical and chemical phenomena caused by them are completely different. The particles also cause different physical and chemical phenomena depending on their energy. Figure 1.2 summarizes that the relations of physical and chemical phenomena caused by particles and photons or quantum particles with the energy transferred/or exchanged and time duration for each phenomenon. The most important point is that the larger the transferred/exchanged energy in the phenomena or the energy of particles, the shorter is the time duration for the phenomena.

Materials consist of atoms and molecules. An atom consists of electrons, neutrons, and protons with the same number of the electrons, and the number is referred to as an atomic number. Atoms are bound to molecules by the interaction of the valence electrons or electrons in the outermost shell of the atoms, and molecules make molecular crystals with intermolecular forces. Many atoms are combined to be a metal sharing their valence electrons. Organic materials consist of polymers and

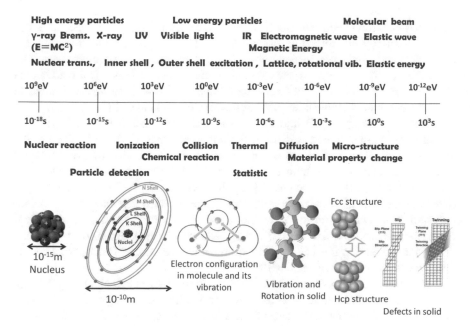

High energy particles   Low energy particles   Molecular beam

γ-ray Brems. X-ray   UV   Visible light   IR   Electromagnetic wave Elastic wave
(E=MC²)   Magnetic Energy

Nuclear trans., Inner shell , Outer shell excitation , Lattice, rotational vib. Elastic energy

| 10^9eV | 10^6eV | 10^3eV | 10^0eV | 10^{-3}eV | 10^{-6}eV | 10^{-9}eV | 10^{-12}eV |

| 10^{-18}s | 10^{-15}s | 10^{-12}s | 10^{-9}s | 10^{-6}s | 10^{-3}s | 10^0s | 10^3s |

Nuclear reaction   Ionization   Collision   Thermal   Diffusion   Micro-structure
                        Chemical reaction   Material property change

     Particle detection   Statistic

10^{-15}m
Nucleus

                                   Electron configuration
                                   in molecule and its
                                   vibration

                                                        Vibration and
                                                        Rotation in solid

Fcc structure

                                                                    Slip   Twinning

Hcp structure

                                                              Defects in solid

        10^{-10}m

Fig. 1.2 Physical and chemical processes, and accompanying energy transfer and time duration

their complex mixing, and living beings are made up of more complex molecular structures. DNA in cells, which is the essential unit for life's support and proliferation having a double helix structure (see Fig. 4.10), is fragile and easily damaged when such small energy as a few eV is given to some of its constituent molecules resulting in the death of the cell. Since EQ are quite tiny and are usually injected discretely, EQ hardly collide directly with DNA in the cell. Unless EQ exposure was not such high dose over a lethal dose, tissues should not be killed by damage of DNA in the cell.

The nuclear energy that combines protons and neutrons in a nucleus is around 10^8–10^6 eV. When a nucleus breaks as nuclear fission or nuclei adhere to nuclear fusion, EQ are ejected carrying energy released by the fission or fusion. Radioactive isotopes (RIs) are nuclei having excess energy in their nucleus, which is released with their disintegration emitting either α- and β-particles or γ-photons as EQ. Since the energy of such EQ is very high (but not high enough to destroy a nucleus), EQ collides with electrons and atoms resulting in ionization and displacement of atoms and molecules in materials. The binding energy of electrons in an atom ranges very widely form the maximum of around 0.1 MeV (100 keV) to the minimum of a few eV. For an atom having large atomic number, the positive charge of its nucleus is so large to bind electrons strongly. Consequently, for ex., the binding energy of the most strongly bound electron in uranium is over 0.1 MeV. Since the binding energy in molecules is mostly less than around 10 eV, EQ can easily break the chemical bond. Accordingly, EQ injected in a material repeatedly collides with electrons and atoms lose their energy by electron excitation, atomic displacements, and breaking molecular bonds until their energy is reduced to around 0.1 eV. It should be reminded that energy transferred/exchanged between atoms above a few eV results in a chemical reaction, EQ having energy of above a few eV could induce unwanted chemical reactions in cells and are hazardous. Therefore, "EQ (Radiation)" discussed here is limited to those having energy above a few eV, irrespective of the kind of EQ.

When the energy of EQ injected in a material becomes less than 1 eV after repeating above-mentioned collisions, the remaining energy of EQ is given to the vibration and rotation of atoms in the materials. Since the vibration and rotation of molecules and atoms exhibit the temperature of the material, all energy of EQ deposited (absorbed) in the material is finally converted to the heat to increase the temperature of the material (thermalization). Some electrons generated by the ionization during the energy transfer processes have enough energy to ionize atoms and repeat the ionization until losing their energy and the energy is also converted to heat. As mentioned earlier, if all constituent atoms of the material got energy of 1 eV, its temperature would increase about 10,000 °C.

In other words, when EQ is injected into a material, they give their energy through various physical and chemical processes, and finally their energy is converted to heat resulting the temperature rise of the material. Figure 1.2 indicates the names of physical and chemical phenomena in these energy conversion processes with indication of their space size and the time duration. During the energy conversion from high to low energy, the space size of the processes becomes larger as the time duration becomes longer. Therefore, the energy and time are simultaneously given in the horizontal axis in Fig. 1.2. Although all EQ energy injected in a material is converted finally to the heat or its temperature rise, the heated volume (the volume exhibiting the temperature rise) is significantly different depending on the kind of EQ either α-particle, β-particle or γ-photon. This difference appears in the influences or effects of the EQ exposure, which is discussed in Chap. 3.

1.2.3 "EQ (Radiation) Exposure" Means Energy Absorption (Deposition) or Energy Transfer from EQ to a Substance

As described in the previous section, EQ exposure to a substance means that the substance is exposed to large number of EQ and some or all of EQ energy is deposited or absorbed in the substance. Depending on the amount of the energy of each energetic quantum and the number of EQ (intensity), the energy absorption (deposition) processes appearing in the substance is largely different. This is one of the reasons why the radiation is difficult to understand.

As an effect of the exposure to ultraviolet light, well known is skin burns which occur when the energy of the ultraviolet light is given to the skin and converted to heat resulting in its temperature rise sufficiently to kill skin cells. This also happens when EQ with energy above around 10 eV is injected into the skin cells and breaks chemical bonds in them.

The effects of the exposure are not necessarily the same even if the total absorbed (deposited) energy is the same for the exposures with lower intensity (smaller number) and higher energy of EQ and with higher intensity and lower energy of EQ. As a specific example, in the following, a comparison is made between the exposure of 10,000 γ-photons with energy of 1 MeV ($= 1.6 \times 10^{-13}$ J) and that of 10^9 photons of visible light of 1.6 eV. Although both give the same deposited energy of 1.6×10^{-9} J, the effects of their exposures to the human body are completely different, i.e., for the former, the effect will be observable, while for the latter, no appreciable effects will appear. Thus, in the discussion of the effects of EQ exposure, not only the total absorbed (deposited) energy (referred to as absorbed dose as explained later), but also both the intensity (the number of EQ exposed) and the energy of EQ matter as shown above.

Another important factor is exposure time or energy absorbed in unit time (energy absorption rate, or absorption dose rate). Consider skin burning by sunbathing for example. The skin burns are mostly caused by the exposure to sunlight. Visible sunlight hardly includes photons having enough energy to give the skin burns for short time of exposure. However, for long time exposure, the integrated number of photons of the shorter wavelength, UV light, or the total deposited energy increases with time resulting the skin burns. Similarly, even for infrared light (photons) with much lower energy than the UV light, long time exposure could cause the skin burns.

In general, photons with lower energy than far-infrared light (microwaves, radio waves, etc.) have no effects on the living beings unless their intensity is very high. On the other hand, low-intensity irradiation of higher energy photons could give some influence, because EQ deposit their energy in quite tiny area/volume, i.e., volume density of the deposited energy (the deposited energy divided by the volume where the energy deposited) is quite large for the EQ exposure, details of which are discussed in the next section and Sect. 4.4 in Chap. 4 again.

In summary, there is three important factors in consideration of the effects of EQ exposure on living beings. They are energy and intensity of EQ, and exposure time. Furthermore, the different kind of EQ gives significantly different effects including the difference between particles and photos. These points are explained quantitatively in Sect. 1.2.6.

1.2.4 Absorbed or Deposited Energy in Unit Mass or Volume is Quite Different Depending on the Kind of EQ

EQ exposure, which is one of the main themes of this book, means absorption (deposition) of all or part of EQ energy to a substance. Therefore, the absorbed energy in the substance under the EQ exposure is one of the most important indicators to discuss the effects of the EQ exposure. The absorbed energy should be normalized by either mass or volume of the substance. The normalization by the mass is physically meaningful and is represented using the units of J kg^{-1}. As already mentioned in Sect. 1.1, in scientific fields related to radiation, a particular unit, **Gray** (G) with 1 G = 1 J kg^{-1} is employed and referred to as **absorbed dose** instead of the absorbed energy. And the absorbed dose in unit time is defined as **absorbed dose rate** or simply **dose rate** and present as Gy s^{-1}, Gy h^{-1}, Gy y^{-1}, etc. depending on the time scale, second, hour, year, etc., respectively. It should be noted that the absorbed dose rate, Gy s^{-1}, is equivalent to W/kg because Gy s^{-1} = J kg^{-1} s^{-1} = W kg^{-1}, i.e., absorbed power in 1 kg. The readers should distinguish the absorbed dose which is absorbed energy in 1 kg and the absorbed dose rate which is absorbed power in 1 kg.

As mentioned several times, the absorbed dose in living beings under EQ exposure varies greatly depending on the kinds of EQ and the species of living beings or organs. In order to normalize the differences in examination of the effect of the EQ exposure on a human body, two weighting factors are introduced: one is the **radiation weighting factor** (W_R) for normalization of the different kinds of EQ, and the other, the **tissue weighting factor** (W_T), for the different tissues or organs as given in Tables 2.1 and 2.2 in Chap. 2, respectively. The absorbed dose (Gy) is converted to **absorbed dose equivalent** defined as **Sievert** (Sv) with multiplication of W_R, i.e., $W_R \times 1$ Gy = W_R Sv. Then the effects of the EQ exposure can be examined irrespective of the types of EQ. The tissue weighting factor (W_T) is introduced for each body tissue or organ to correct the difference in the effects of exposure among them. Thus, converted dose equivalent (the absorbed dose equivalent multiplied with W_T) is referred to as **effective dose**. Since the effect of EQ exposure is discussed based on these four parameters, Gy, Sv, W_R, and W_T, understanding these parameters or what they mean is indispensable as described in following sections and chapters.

In most cases, public exposure is caused by β-particles and γ-photons. For both the values of W_R is nearly the same and accordingly absorbed dose equivalent given in Sv is nearly the same as Gy. While W_R of other EQ, α-particles, neutrons and heavy ions are quite different from those of β-particles and γ-photons. Therefore, the comparison of the effects of exposure caused by the different types of EQ based on Sv should be made carefully.

1.2.5 Units Related to Radiation, Exposure, and Radiation Measurements

1.2.5.1 Energy and Power Carried/Deposited by EQ (Radiation) (J or eV and W)

Usually, radiation measurement means to detect and count EQ coming into an instrument named a radiation counter. In case that an instrument can measure energy carried by EQ and deposited in it, it is referred as a dosimeter. For full accounting of EQ (radiation), it is necessary to determine the kind of EQ, and their energy distribution, because EQ generally includes different types and each type of EQ consist of quanta with different energy, i.e., EQ shows energy distribution. The integration of the measured energy distribution gives the absorbed or deposited energy in the instrument, which can be converted to absorbed dose in a substance placed at the location of the instrument.

Here reconsider physical parameters relating EQ (radiation), intensity, energy, kind of EQ, and EQ sources. EQ carries energy, as already mentioned, the energy that individual quantum has is represented by electron volts (eV) or collectively by joules (J). And when a quantum collides with a substance, all or some of its energy is absorbed (or deposited) in it. Since the absorbed energy is evaluated per unit time, it is expressed as the absorbed energy per unit time, i.e., $J s^{-1}$, which is just the power in physics and is defined as watt (W). However, when dealing with EQ, the expression "power" is usually not used. Nevertheless, there is no difference in absorbed or deposited energy and power used in EQ exposure and energy and power in electricity for example. Thus, absorbed dose expressed by Gy ($J kg^{-1}$) and absorbed dose rate, Gy s^{-1} (W kg^{-1}) are equivalent to, respectively energy absorbed or deposited, and power deposited in unit weight of the substance.

EQ carrying energy of above 10 eV is scary, so as EQ emitted from radioactive materials in nature whose energy is mostly above 0.1 MeV (1 MeV = 10^6 eV) or 10^{-14} J. When the intensity of EQ in nature is measured by a simple radiation counter, it would not exceed 10 counts per second. That is, the number of EQ coming into the counter is less than 10 per second or roughly 100 per minute. Therefore, the power given by EQ in nature is about 10^{-14} (J) × 10 (s^{-1}) = 10^{-13} W or higher. Usually, an infrared heater radiates a power of several W per 1 cm^2. So, if a person is 10 cm away from the heater, he will receive the power of approximately 10^{-2} W cm^{-2} or 10^2 W m^{-2}. Compared with the power given by the heater, the power given by natural EQ (10^{-13} W) is nearly nothing and the readers would wonder why EQ is dangerous.

To give the answer to this wondering, the key is on how large area/volume the EQ energy is absorbed. The size of an energy quantum is less than 1 nm (1 nm = 10^{-9} m). Suppose a quantum with an energy of around 1 MeV or 10^{-14} J enters a circle with a radius of 1 nm and gives the all energy, then the power deposited in the circled area (3 × 10^{-18} m^2) becomes 3 × 10^4 W m^{-2}. Compared to the power given by the above-mentioned infrared heater, 10^2 W m^{-2}, the power given by one quantum is more than 100 times larger but on very tiny area, 3 × 10^{-18} m^2. This is one of the main reasons why EQ exposure is dangerous. If the energy of individual quantum was deposited in wider area, the power would be quite small and nothing would happen on the exposure. Instead, EQ exposure gives quite large power in quite small areas and could cause the direct destruction of the cell in the living beings exposed to EQ.

However, as a matter of fact, if a human body was exposed to EQ with 100 cps per m^2 (this is approximately equivalent to 1 micro sievert (μSv) in the air dose rate), the effect would appear in a very tiny area of 100 × 3 × 10^{-18} = 3 × 10^{-16} m^2. Even if the dose rate was 10^6 times larger, i.e., 1 Sv, with which exposure some effects should appear in a human body, the actual exposed area is 3 × 10^{-10} m^2, only about

20 μm in diameter. Nevertheless, exposure to EQ deposits energy volumetrically, so that it is likely the penetration of a needle into the body with the radius of about 20 μm. The needle gives pain, while the exposure of EQ does not. That is exactly why EQ is invisible and insensible. More details are given in Sect. 4.4 in Chap. 4.

From above discussion, one can understand that EQ exposure does not occur uniformly such appears in the exposure to air and water, but a huge power or energy is deposited quite locally in a very tiny area. Therefore, the influence of the EQ exposure appears differently depending on where EQ hit or enters. If they entered an important organ in a human body, it would give significant effect like a cancer. On the other hand, if they hit the part of limbs, it would not result in significant damage. In case of lower dose exposure, the region or point of a human body where EQ hit is quite statistical, and the influence appears quite differently depending on the region. As absorbed dose rate increases, statistical scattering for the appearance of the effect of EQ exposure becomes less. Accordingly, the higher dose of the EQ exposure, above about 1 Sv, anyone will obviously be affected.

All above estimations were done by so to speak an order estimation, for easier understanding. In addition, as mentioned several times, the absorption dose rate varies depending on the type of EQ, their energy, and exposed substances, and so on, which give additional uncertainty to the estimation. Therefore, in the discussion of appearance of some irradiation effects, the estimated absorbed dose-dependence could include the inaccuracies of more than one digit.

1.2.5.2   Absorbed Dose and Dose Rate

The main object of this book is to help the readers to understand what is radiation or in more detail how EQ energy (radiation energy) is deposited or absorbed in a substance exposed to EQ. At the EQ exposure, all energy of EQ is not necessarily transferred to absorbed energy in the substance. In addition, the way how to transfer the energy to the substance varies greatly depending on the type of EQ, their energy, and the nature or property of the substance. Therefore, the absorbed energy per unit weight of the human beings or living being is a measure of the EQ exposure as introduced as the absorption dose ($1 \text{ Gy} = 1 \text{ J kg}^{-1}$).

As also mentioned, since the absorption dose in a human body varies greatly depending on the type of EQ and its energy, the radiation weighting function (W_R) is introduced for discussion of irradiation effect with the exposure to different types of EQ, and the absorbed dose (Gy) is converted to the absorbed dose equivalent (Sv). In addition, since the manner or amount of energy absorbed in the human body varies depending on character/nature of tissues or organs, an additional weighting factor, i.e., the tissue weighting factor (W_T), is introduced for each tissue or organ to correct the different nature among them and the absorbed dose equivalent is converted to the effective dose. These points are explained in detail in Chap. 2.

The author is concerned that it would enhance the scariness of EQ exposure or radiation to have introduced unique parameters such as the absorption dose (Gy) and the absorbed dose equivalent (Sv) instead of the universal unit of J/kg for discussion of effects of EQ exposure. There is additional concern to give misunderstanding of the radiation as something special from the conversion of the absorption dose equivalent to the effective dose for discussion of the exposure effects on different tissues or organs in a human body. Although such units have been introduced with necessity and traditionally used in research of radiobiology in its long history, the use of such units particularly the absorbed dose equivalent (Sv) could easily lose the physical meaning or truth in the radiation effects appeared as the results of energy absorption or power deposition, which is the essence for discussion of exposure effects of EQ. In this respect, it would be better to use J kg^{-1} or Gy under clear indication of the type of EQ and the type of tissue. A simple radiation detector gives only the intensity of EQ as counting rate like cps or cpm, while the energy distribution cannot be determined. A handy dosimeter is often used, in which the total deposited energy is measured and the results are presented with Sv or Sv t^{-1} with a simple assumption that EQ consists of β-photons or γ-photons. Hopefully, the discussion of the radiation effects would be done with correct understanding of Sv. More detail of this point is discussed in Chaps. 2 and 6.

1.2.5.3 Intensity of EQ or Radioactivity

Following the discussion of the energy of EQ, the intensity of EQ (radioactivity) is discussed in this section. The intensity of EQ, or the number of EQ emitted or received, is measured by a simple radiation counter as cps, cpm of cph. Radiation measurements include determination of the type of EQ, their intensity, and energy distributions of EQ both for EQ sources and for an exposed substance. For the exposed substance, it is necessary to determine how much energy is absorbed or deposited. The intensity of EQ emitted from the source consisting of RI is referred to as the radioactivity of the source given by Bq (the number of EQ emitted from its source in one second), or dps (disintegration per second), dpm (per minute) or dph (per hour), while the number of EQ detected is expressed as cps (counts per second), cpm (per minute), or cph (per hour), which is exactly how many EQ the instrument detected per second, minute or hour.

In general, since EQ is emitted in all directions from its source, it is necessary to consider the geometrical factor of an EQ detector and an EQ source to determine the total number of emitted EQ from the source, i.e., the radioactivity of the source. If the EQ source is a radioactive material like RI, EQ is radiated evenly in all directions, while if the source is an artificial radiation generator or multiple sources, the emission of EQ is not uniform but directional. Then the intensity is represented as flux, i.e. the number of EQ passing through or injected into a unit area, i.e., cps m^{-2} or Bq m^{-2}.

The air dose, which is the absorbed dose given by EQ sources in nature, is determined as the absorption energy in a human body usually a particular area or volume of a human phantom (a model of the human body) exposed to EQ. Hence it significantly changes with geometrical relation between the source and the body. In a simple dosimeter, its absorption energy given by EQ exposure is measured and converted to the absorption energy in a human body (differences among tissues are averaged out), and then the converted value is indicated as the air dose equivalent or effective dose with the Sv unit. To get the correct absorbed dose, the geometry of the source and intensity of EQ, and the structure of substance should be considered. This point is revisited in Sect. 6.2.5 in Chap. 6.

1.2.6 Intensity and Energy of EQ

"Intensity" and "energy" of EQ are often confused in discussions of the effect of EQ exposure. The expressions like "large and small", "high and low", and "strong and weak" are also used to qualify EQ or EQ sources in confusion. People who have studied physics are used to the term "high energy physics", so one can immediately see that high and low energy corresponds to large and low the amount of energy. However, they are also used to express large and small numbers, respectively as synonyms of high and low. In addition, "large" or "high" and "strong" may be used as synonyms in Japanese, and "small" or "low" and "weak" may be used as synonyms, which may deepen confusion. When dealing with EQ (radiation), the energy of EQ carrying is referred high and low, the amount of absorbed energy (absorbed dose), large and small, while the intensity of EQ is distinguished using the expression "strong or intense" and "weak", but this terminology does not seem to be understood correctly.

In this book is explained that (1) radiation is composed of EQ carrying energy, and (2) exposure to EQ results in energy absorption in a substance exposed to EQ. Therefore, particular focus is given to the intensity and the energy of EQ. However, it may be difficult to understand without using clear definition for both with using mathematical formulas (equations) and quantitative expressions. All physical parameters required for discussion of the EQ exposure are introduced in previous sections. Therefore, this section is devoted to explaining quantitatively the intensity and energy of EQ (radiation) using formulas, although the subjects are overlap with those given in previous sections.

Radiation effects are some changes caused by the EQ exposure of a substance through the energy absorption in it. Since the details of energy absorption processes of EQ penetrating into the substance are discussed in Chap. 2, only the initial process of absorption of energy or power of EQ injected in the substance is explained for easy understanding of previous section, i.e., "high and low" and "large and small" are used to express EQ energy, and "high (intense) and low" to the intensity of EQ.

As described in Eqs. (1.1) and (1.2), the total energy of EQ carrying, E, is the sum of kinetic energy of each particle,

$$E = \sum_{i,j} \frac{1}{2} m_i v_j^2 \tag{1.7}$$

where i and j stand on different mass and velocity, or the sum of each photon energy

$$E = \sum_j h v_j. \tag{1.8}$$

where j stands on different frequencies. The energy carried by EQ increases with increase of the velocity and frequency. It also increases with the number of particles given by i and j, which is referred to as the intensity or flux. Therefore, high/low energy EQ and high/weak intensity of EQ can be clearly distinguished. However, the total energy carried by EQ which is compared with large or small does not indicate whether EQ consists of high/low energy particles or photons, or the intensity of EQ is high in numbers of particles and photons.

When the type of EQ is identified, the energy they carry is simply given by

$$\varepsilon_j = \frac{1}{2} m v_j^2 \tag{1.9}$$

or

$$\varepsilon_i = h v_i. \tag{1.10}$$

The next important point is whether (a) the source of EQ is distributed/dispersed in space or (b) it has a limited volume, as separately discussed in the following.

(a)   EQ sources distributed in space

Suppose EQ are distributed in space with each quantum having energy ε with density $n(\varepsilon)$ (particles m^{-3} J^{-1}), the total energy of EQ, E, is given by

$$E = \int \varepsilon \times n(\varepsilon) d\varepsilon \ (J\ m^{-3}) \tag{1.11}$$

It should be noted that the dimension of $n(\varepsilon)$ is particles m^{-3} J^{-1}. Energy distribution indicates the numbers of quanta as the function of their energy under the condition of the total number of quanta, N_0, is constant (limited), i.e.,

$$N_0 = \int n(\varepsilon) d\varepsilon. \tag{1.12}$$

Hence the dimension of $n(\varepsilon)$ is particles m^{-3} J^{-1}.

If EQ sources are uniformly distributed in the space around a substance, the energy given by Eq. (1.5) is deposited into the substance. Then dividing the energy by the density of the substance (kg m^{-3}), one can get the deposited energy or absorbed dose in the substance with the unit of Gy (J kg^{-1}). Of course, since all energy is not necessarily absorbed in the substance, the absorbed dose could be somewhat different from the deposited dose. Therefore using a dosimeter, described in Chap. 6, the absorbed dose is determined from the deposited energy in it. Since the manner of energy absorption is different depending on the kind of EQ, a correction factor

for different kinds of EQ, referred as a radiation weighting factor (W_R) is introduced. Then the absorbed dose is converted to the absorbed dose equivalent (Sv).
Furthermore, the effect of the EQ exposure is also different depending on tissues or
organs. Hence another weighting factor, a tissue weighting factor, W_T is introduced
to normalize the differences among them, which give effective dose also represented
in Sv. Considering the measuring time, absorbed dose rates, Gy s^{-1}, Gy m^{-1}, or
Gy h^{-1} are given so as the dose equivalent rates, Sv s^{-1}, Sv m^{-1} or Sv h^{-1}.

Note that a normally used dosimeter gives the absorbed dose equivalent in Sv as the
integration of its dose rate for measuring duration, while a simple radiation detector
(counter) usually gives the intensity or the number of detected EQ in specified time
duration, cps, cpm or cph.

(b)   EQ sources limited in volume

Natural air dose is given by EQ sources almost uniformly distributed in atmosphere,
while unintentional EQ exposures, like the nuclear accident, is caused mostly by
localized sources with limited volumes. In the latter case, it is necessary to consider
the direction of EQ injecting into a substance. In other words, to examine effects of
EQ exposure on the substance, particularly a human body, caused by the localized
source, it is necessary to know areal and volume distributions of deposited energy
of the incident EQ, i.e., the power flux (J m^{-2} s^{-1}) and absorbed energy density
(J m^{-3} s^{-1}) on the substance given by the EQ exposure.

When EQ is emitted from a limited volume source, different from $n(\varepsilon)$ where the
sources are distributed in space, flux of EQ, $\phi(\varepsilon)$, i.e., the number of EQ with energy
of ε injecting to the substance in unit area and unit time (particles m^{-2} s^{-1} J^{-1}) is
accounted.

The total number of EQ injected into a substance in unit area is given by

$$\Phi_0 = \int \phi(\varepsilon) d\varepsilon, \qquad (1.13)$$

which corresponds to the intensity of exposed EQ given as Bq m^{-2}. The total injected
energy in unit area and time is given by,

$$P = \int \varepsilon \times \phi(\varepsilon) d\varepsilon \qquad (1.14)$$

with the unit of J m^{-2} s^{-1} which is the same as the total deposited power in unit area,
W m^{-2}. In the simplest case, the exposure to EQ all having the same energy of ε_0
with the total number of N_0 gives the energy of $\varepsilon_0 \cdot N_0$ in J m^{-2} s^{-1} in unit time or
the power of $\varepsilon_0 \cdot N_0$ in W m^{-2}.

Thus, EQ exposure to a substance means that the substance is exposed to power
(or EQ deposit the power to the object) and its time integration becomes deposited
energy. And the power and energy absorbed in the substance are accounted for as
absorbed dose rate in Gy t^{-1} and its time integration results in the absorbed dose in
Gy.

When EQ energy is high or the EQ intensity (incident EQ flux) is high, the absorbed dose becomes large. However, the effects of the EQ exposure are not the same for the two cases even giving the same deposited energy; (i) EQ has higher energy with lower intensity, and (ii) EQ has lower energy with higher intensity. Hopefully, this makes the readers understand the absorbed energy and intensity of EQ (radioactivity).

Thus, EQ exposure means that the exposure to the power of EQ and the time integral of the power becomes the absorbed dose. And as depicted in Eq. (1.14), both the energy and the intensity of EQ contribute to the absorbed dose.

For the consideration of EQ exposure, the absorbed dose or integrated power with time is not enough, i.e., the kind, intensity, and energy of EQ should be simultaneously considered. Even for the same absorbed dose, one of the three is different, the effect of exposure could be significantly different. Although the difference is considered by employing the radiation weighting factor (W_R) and the tissue weighting factor (W_T), that is not enough, too.

The following summarizes the above discussion. At first, the intensity of EQ or the EQ source is represented as the number of EQ injected in unit area given in cps or emitted from the source in unit time given in Bq which is referred to as radioactivity of the EQ source. Then the energy of EQ carrying or emitted from the source is considered. Assuming a point EQ source, which emits N_{Source} (Bq) of EQ. (Although the EQ emission from RI decreases with its half-life time, it is nearly constant in a short time.) The energy ε of each EQ is not necessarily the same, but has energy distribution given by $n(\varepsilon)$ relating to the total emission N_{Source} as

$$N_{Source} = \int n_{Source}(\varepsilon) d\varepsilon, \qquad (1.15)$$

according to Eq. (1.12). Then the total energy carried by EQ or emitted from the EQ source is given by

$$E_{Total} = \int \varepsilon \times n_{Source}(\varepsilon) d\varepsilon \qquad (1.16)$$

according to Eq. (1.11).

On the other hand, the intensity of EQ received by a substance exposed to EQ is also given in Bq. Following above discussion, the intensity of EQ the substance-exposed, N_{Target} (Bq), is given by

$$N_{Target} = \int \Phi_{Target} ds = \int \int \int \phi_{target}(\varepsilon) d\varepsilon dt ds, \qquad (1.17)$$

where Φ_{Target} is the total flux of EQ and $\phi_{target}(\varepsilon)$ is the flux of EQ having energy of ε. The integration is done with energy (ε), exposed area (s) and time (t). Then the power P_{Target} loaded to the substance is given by

$$P_{\text{Target}} = \int \varepsilon \times \phi_{\text{Target}}(\varepsilon) \mathrm{d}\varepsilon. \tag{1.18}$$

Since all power loaded is not necessarily absorbed in the substance, the absorbed power, P_{Absorbed} in the object can be written as

$$P_{\text{Absorbed}} = k \times P_{\text{Target}}, \tag{1.19}$$

where k is an absorption coefficient. If the mass of the object is M (kg), the absorbed dose rate becomes $P_{\text{Absorbed}} \cdot k^{-1} \cdot M^{-1}$ (Gy s^{-1}). Table 1.2 summarizes EQ sources, and intensity and energy emitted from them, while Table 1.3, energy deposition (absorption) given by EQ exposure.

An additional consideration is required on the relationship between the emitted power from the source and the loaded power to the object. Since there is a space between the EQ source and the substance (unless both are in close contact) and both have geometrical structures, the power given to the substance varies depending on their distance and geometrical relationship. Since part of the EQ energy is absorbed in a material filling the space, it also influences the deposited energy, or the absorption coefficients of both the material filing the space and the substance itself are important factors. Considering all these points, the conversion from the deposited energy to the absorbed energy or absorbed dose, and even to the dose equivalent should be

Table 1.2 Summary of radiation sources

Energetic quanta (EQ)		Sources in atmosphere
α-particle β$^-$-particle (electron) β$^+$-particle (positron) γ-photon (γ-ray) X-ray, soft X-ray Neutron Elementary particles (Pion, Muon, etc.) Electron beam Ion beam	Particle energy $\varepsilon = \frac{1}{2}mv^2$ Photon energy $\varepsilon = h\nu = \frac{ch}{\lambda}$	Radioisotopes dispersed in air and water, and on-ground surface Solar radiation (soft-X-ray, UV light, visible light) Cosmic ray
Planer source $\varphi(\varepsilon)$ (EQ flux)	Flux $\Phi = \int \varphi(\varepsilon) \mathrm{d}\varepsilon$ (Bq m^{-2} s^{-1})	cps, Bq
Volume source $n(\varepsilon)$ (EQ intensity)	Intensity $N_0 = \int n(\varepsilon) \mathrm{d}\varepsilon$ (Bq kg^{-1} or Bq m^{-3})	cps, Bq
Emitted energy	$E = \int P \mathrm{d}t$ $= \iint \varepsilon \times \varphi(\varepsilon) \partial\varepsilon \partial t$ (J or eV)	Absorbed energy Air dose (Gy; 1 Gy = 1 J kg^{-1})
Emitted power	$P = \int \varepsilon \times \varphi(\varepsilon) \mathrm{d}\varepsilon$ (J s^{-1} or W)	Air dose rate (Gy s^{-1})

Table 1.3 Summary of energy absorption (deposition) given by EQ exposure

Exposure to substance (energy absorption (deposition))			
Exposed flux	$\Phi = \int \varphi(\varepsilon)d\varepsilon$		Bq m^{-2} s^{-1}, cps m^{-2} s^{-1}
Exposed intensity	$N_0 = \int n(\varepsilon)d\varepsilon$		Bq kg^{-1} or Bq m^{-3}
Power flux	$P = \int \varepsilon \times \varphi(\varepsilon)d\varepsilon$		W m^{-2} = J s^{-1} m^{-2}
Absorbed energy[a] or deposited energy	$E = \int Pdt$[a]	Absorbed dose	Gy (1 Gy = 1 J kg^{-1})[a]
		Dose equivalent	Sv = W_R Gy W_R: radiation weighting factor for different kinds of EQ (see Table 2.1)
		Effective dose	Sv = $W_R \cdot W_T$ Gy W_T: tissue weighting factor for different tissues (see Table 2.2)

[a]To discuss the effect of exposure, both exposure time and volume or weight of an object exposed should be identified

done. The geometrical relationship between the source and the exposed substance is discussed in detail in Sect. 2.3 in Chap. 2.

1.3 Energy Release from Materials (Black Body Radiation and EQ Emission)

Historically, in 1859, Gustav Kirchhoff has discovered that even a black material, which does not emit visible light at room temperature, radiates infrared and visible light at high temperatures. In 1900, Max Planck has given a theoretical equation showing the temperature dependence of the wavelength and intensity of the emitted light (Planck distribution). This is called black body radiation as shown in Fig. 1.3. Radiation power $p(\lambda)$ from a black material at temperature T is given by the Plank's equation,

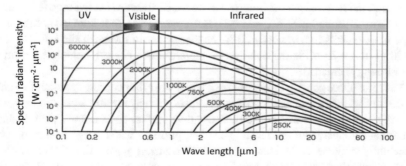

Fig. 1.3 Black body radiation; changes of wavelength distribution with temperature

$$p(\lambda) = \frac{8\pi hc^2}{\lambda^5} \frac{1}{\exp(hc/\lambda kT) - 1},$$ (1.20)

where λ, h, and k are the wavelength of the radiated electromagnetic wave (light) in m, the Plank constant in J s, and the Boltzmann constant in J K^{-1}, respectively. As seen in Fig. 1.3, up to about 5000 °C, the wavelength of the emitted light is continuously distributed in wide wavelength regions, and the higher the temperature of the material, the shorter the wavelength of the emitted light. Although in actual materials, the temperature dependence of the radiation power varies with their character, the black body radiation appears above 0 K. Measuring the radiation from a human body, the temperate of people entering a country at an international airport can be measured, so that those who show high body temperatures (may be sick) can be checked.

The fact that a person's body temperature is kept constant is caused by the power balance between the input and output, i.e., the sum of the power generated in the body by the combustion of carbohydrates (called metabolism) and the power incident from the outside (like an exposure to the sun light) is balanced with the total radiation power from the body. Hard exercise increases the metabolism (the larger energy release) and increases the body temperature which is compensated by sweating. If this balance is broken and cooling is insufficient, one could get heat stroke.

The earth receives energy from the sun and at the same time radiates energy into space. It should be noted that the blue color of the earth seen by an astronaut (hence the earth is called "Blue planet") does not mean the radiation from the earth is blue. The reason for the blue-colored earth is due to the reflection of blue part of the sunlight but not due to radiation from the earth's surface. The radiation from the earth is mostly consisting of infrared and far-infrared light nearly equivalent to those radiated from a material at around -23 °C. If there was no radiation, the earth would warm up to higher temperature. The cause of global warming is that some of the radiation from the earth cannot be escaped due to the greenhouse gases.

Figure 1.4, shows energy balance of the earth. The left of Fig. 1.4a shows the wavelength (energy) distribution of the radiation of the sun and the radiation from the black body at 5525 K. The right of Fig. 1.4a is the wavelength distribution of the radiation from the earth and the black body radiations of 210–310 K indicated by pink, blue and green. The power balance between the incident power (energy) from the sun and the radiated power (energy) from the earth keeps the average temperature of the earth at around 250 K (-23 °C). (Of course, it is well known that there is a temperature distribution of about 20 °C on the surface of the earth and -40 °C above 10,000 m, but -23 °C is the average temperature of the earth calculated from the power (energy) balance.)

The atmosphere contains various gases which absorb specified wavelength light as detailed in Fig. 3.6 (see Chap. 3). Thus, the absorption by the atmosphere changes the wavelength distribution that arrives at the earth's surface so as the wavelength distribution that radiates from the earth. The absorption by water, oxygen, and nitrogen are natural, while recently increasing global warming gases, like CO_2 absorb the radiation from the earth, mainly in infrared region corresponding the global warming.

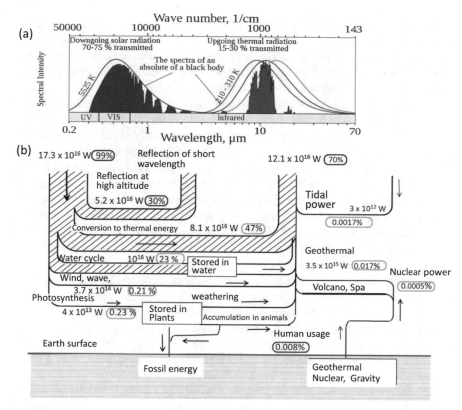

Fig. 1.4 a Radiation spectra from the sun and the earth (https://en.wikipedia.org/wiki/User:Dra gons_flight/Images#/media/File:Atmospheric_Transmission.png, drawn by Dr. Robert A. Rohde, reused with permission). **b** Power balance of the earth

Above around 10,000 °C, the wavelength of the emitted light becomes very short, and the invisible ultraviolet light increases, and at the same time, the wavelength distribution becomes no longer continuous and deviates from the black body radiation. Moreover, the wavelength distribution is different depending on materials. At higher temperatures, electrons bound in atoms are released, or atoms are ionized to be plasma state. In the plasma, all particles have a large energy, and if they escape from the plasma, the escaping particles and light from the plasma are the so-called "radiation" by themselves. Therefore, radiation is the emission of energy from materials at high temperatures or in high-energy states.

There is no difference in basic physics of the energy balance of the earth and that happened in a substance exposed to EQ. As described several times, the EQ exposure results in the absorption of all or part of EQ energy injected into the substance. The absorbed energy in the substance finally becomes thermal energy and raises its temperature. For human beings, the temperature rise is compensated by radiation and sweating to keep the temperature constant. However, the absorption energy given by EQ exposure is normally too small to be measured as temperature increases. In case of EQ, which has much higher energy than ultraviolet or visible light from the sun, energy conversion process includes more physical/chemical processes than that appeared on the exposure to the sunlight before fully thermalized as seen in Fig. 1.2, and these processes contribute the appearance of radiation effects.

Since an intense EQ source emits large amount of energy, if it is thermally isolated from its surroundings, the emitted energy is converted to heat resulting in the temperature rise of the source and its container (self-heating). Actually, spent fuels of a nuclear reactor, which contain various radioactive fission products, are kept in water pool to shield emitted EQ and remove the released heat. In the accident of the Fukushima nuclear power plant, loss of the cooling water in storing pools of the spent fuels was also the major concern. On the other hand, heat released from RIs can be used for generation of electricity by thermoelectric conversion. In space satellites, radiation batteries in which the thermoelectric conversion of released energy from RIs are used for the generation of electricity. Different from other power sources, the radiation battery continues power supply without oxygen (combustion).

Most of elements in the periodic table which are expressed as $_A^m Z$ with m, A, and Z are, respectively the mass number, the atomic number, and the name of an element, have isotopes, which are the same in A but different in m. Several elements have a radioisotope (RI) which has excess energy in its nucleus and the excess energy is released as EQ. Since the excess energy is finite, the energy release exponentially decreases with time showing its own half lifetime and finally, it becomes a stable isotope.

Although most of EQ in nature is given by RI, EQ sources can be artificially created. For example, in a device called a particle accelerator, electrons or ions are accelerated in an electromagnetic field (giving energy) to high energy state or high speed. The accelerated electrons or ions are exactly EQ. In electron microscopes, accelerated electrons are used to make images of an object with electrons reflected or permeated. The accelerated electrons are the same as β-particles. There is no difference between them except the β-particles generally having much high energy. Recent technological advances make it possible to artificially create EQ even a new element with the head-on collision of two accelerated heavy ions into ultra-high speed. One of the examples is the new element called Nihonium ($_{131}^{278}$Nh). Although, electrons and ions can be controlled by electric or magnetic fields (accelerating, decelerating, or changing the direction of their movement), photons (X- and γ-photons) are extremely difficult to control. This is the one of the reasons why EQ is "scary". The effective use of EQ is described in Chap. 7.

1.4   EQ Sources in Nature

The universe is filled with EQ. The earth is a special planet where living beings can live or survive without dangerous EQ. Except for the earth, no planets where any living beings exist have been found. The sun in which energy is generated by nuclear fusion reactions emits the energy as EQ (mostly as photons or electromagnetic waves) from its surface. The sun is the largest and the nearest energy source for human beings. The energy of single energetic quantum is ranging from a maximum of about 10 MeV (10^7 eV) to a minimum of 0 eV. For comparison, the energy of single-photon of radio wave is in the order of 10^{-6} eV.

EQ with energy of more than several eV is hazardous for living beings. Fortunately, very high-energy EQ generated in the sun lose their energy inside and does not come out from the surface. EQ reaching the earth show energy distribution similar to the blackbody radiation at about 5800 K, as described in Sect. 1.3. Since high-energy EQ are also shielded by the atmosphere, only photons with energy ranging from around 5 eV to 1 meV reach the ground surface as visible light (1–5 eV) and infrared light (1 eV to 1 meV) which are invisible but are detected as heat. In addition, as explained in Sect. 1.7, RIs that have existed since the birth of the earth remain inside. It is believed that fission and fusion reactions occur due to high pressures inside the earth, but EQ emitted by them is shielded with the earth's crust.

Although searches for any living beings in the universe are continuing, until now no lives are found on any stars or planets except the earth. People have a dream to travel to the moon or Mars and it is likely realized. In reality, however, such travel requires not only protection from vacuum with a space suit but also radiation shielding. In particular, the travel to Mars depends on how radiation exposure can be reduced. On the earth, there certainly exist natural radiation or EQ sources. Except for several particular areas shown in Table 1.4, the air dose rates (which will be explained later) is well below 1 mSv per year (1 mSv y^{-1}) which ICRP considers a safety guideline not to give major damage.

The earth is basically kept at a nearly constant temperature with a balance between the incident power from the sun and the radiation (heat emission) from the earth. Although global warming has been remarkable in recent days, the earth is gradually cooling down from the birth on a long time scale of over million years (the earth

Table 1.4 Air dose rate in various areas in the world

	Area (country)	Air dose rate (mSv/year)	
		Average	Maximum
High dose rate area	Ramsar (Iran)	10.2	260
	Guarapari (Brazil)	5.5	35
	Karunagappalli (India)	3.8	35
	Yangjiang (China)	3.5	5.4
Comparison among countries	Norway	0.63	10.5
	Hong Kong	0.67	1.0
	China	0.54	3.0
	Germany	0.48	3.8
	Japan	0.43	1.26
	US	0.40	0.88

was born about 4.6 billion years ago). However, the cooling rate is delayed due to the energy release from natural RIs remained in deep inside of the earth. Among various RIs, ^{40}K, ^{87}Rb, ^{147}Sm, ^{176}Lu, and ^{187}Re have been existing since the birth of the earth, because their half-lives are longer than 4.6 billion.

In ancient earth, oxygen in atmosphere was thin and absorption of ultraviolet light in the atmosphere to form ozone did not take place so that on the grand surface the intensity of sunlight in shorter wave length regions was much higher than today and the living beings were difficult to exist. However, owing to the shielding of EQ by water, some primitive lives appeared in seawater and evolved to higher protists. EQ from the space has very likely influenced the evolutionary process of lives. Recently, the famous theory of Charles Darwin is subjected to the criticism that some mutual evolution adopted to the environment survived without any selection rules. Exposure to EQ is considered as one of the causes of the mutual evolution. In Chap. 9, history of the earth and the development of lives are described in a little more detail. It is often claimed that many people instinctively hate or fare reptiles, because human beings were threatened by the reptiles, typically dinosaurs in Jurassic period. Probably fear of EQ is also instinctive like the hate of the snakes.

In the following is considered the magnitude of energy or power from the sun. As shown in Fig. 1.4, basically, the temperature of the earth's surface is kept almost constant because the incident power from the sun mostly visible light, and the radiation form the earth as infrared light is balanced. In outer space of the earth, the energy or power given by the sun in unit area is approximately 340 J s^{-1} m^{-2} or 340 W m^{-2}, respectively. In the atmosphere, higher energy photons or ultra violet light is absorbed. Consequently, the remaining visible light (around 1–5 eV) gives the power of about 240 W m^{-2} to the surface of the earth.

It should be noted that the power given by the sun to human beings on the earth surface is almost comparable to the power that they consume in ordinary life, around 100 W m^{-2}. This is quite natural. If the sun gave much more power, they would be heated up and should be cool down. As long as a person continues consuming energy, the energy is converted finally into heat which must be removed by cooling. Since the cooling power of the human beings is limited so that the power input should not exceed it, otherwise their temperature continuously increases getting heatstroke. Therefore, the power input must not be much higher or lower than their power output by metabolism. In other words, a small difference in input power from outside or temperature of the atmosphere is easily detected as warm or cold feeling.

By the way, a solar cell with an energy conversion efficiency of 10% generates electric power of about 20 W m^{-2}. A generator equipped to a bicycle generates a few W of the electricity which gives enough brightness for LED recently employed. When a human works hard, he can generate about 100 W of power. If the conversion efficiency from work to electricity in the generator of the bicycle is about 10%, then he can generate the electric power of 10 W.

Dangerous EQ can also be a useful energy source if they are converted well under safety control. Nuclear reactors convert the dangerous EQ energy generated by nuclear fission reactions to electricity and so as the radiation battery as described above. The research and development of a fusion reactor that uses the nuclear fusion reactions like the sun have been promoted under the catchphrase of "to build the sun on the ground!" but have not been realized yet.

1.5   Energy Transfer in Physical and Chemical Phenomena

Table 1.1 shows energy carried by electromagnetic waves and corresponding wavelength and frequency over 20 digits. Depending on the amount of their carrying energy, the electromagnetic waves are referred to as different names and used for different purposes. As shown in Fig. 1.2, any physical and chemical processes accompany energy transfer. Furthermore, the time duration occurring each physical or chemical process becomes longer with decreasing energy transferred, because the speed of the particles is slower, while the space or volume and number of particles involved in the processes become larger.

In any way, it is said that the universe started with Big Bang and the energy of about 10^{70} J or 10^{90} eV was released as EQ. Apart from the cosmology, in the sun, our energy source, a nuclear fusion reaction of four protons (p) and two electrons (e^-) generates the energy of around 27 MeV,

$$4p + 2e^- \rightarrow\,^4 He + 6\gamma + 2\nu + 26.65\ MeV, \tag{1.21}$$

where γ and ν are a γ-photon and a muon, respectively (see Fig. 3.5 in Chap. 3). According to Fig. 1.2, the reaction time of the fusion reaction would be about 10^{-17} s. Most of the released energy of 26.65 MeV is distributed to the He ion and γ-photon. When they are injected into a material, electrons bound to the constituent atoms of the material are released or the constituent atoms are ionized. The energy required for the ionization is around several eV or above, and the reaction time is about 10^{-12} s. Electrons and ions thus produced collide with other electrons and ions to produce more electrons and ions having lower energy. These processes are repeated until the energy of the electrons and ions is reduced to an eV order. During the collision processes bonding between atoms or chemical bonds in the material are broken accompanying formation of some new chemical bonds or defects. The time required for these processes is about 10^{-9} s. As such, the energy initially carried by EQ is transferred/distributed to many or various atoms and electros in the material. If this happens in living beings, cells in tissues or organs are damaged resulting in the death of the cells or canceration of the cells and tissues.

When the energy is further transferred to more atoms and molecules to be less than 1 eV, no more chemical reaction is possible but the energy is converted to vibrational and rotational energy of atoms and molecules. Thus, through various physical and chemical processes, 1 MeV of one photon, for example, is finally converted into 10^9 photons of vibrational and rotational motions of atoms in the material with the energy of meV, which appears as the temperature rise of the material. However direct conversion of one 1 MeV photon to 10^9 photons having the energy of 1 meV is impossible. The energy conversion always proceeds step by step passing through energy range causing ionization, electron excitation, vibrational and rotational excitations of molecules, all of which induce radiation effects or damage to living beings.

Here is considered energy release from the sun again. The energy of 26.7 MeV generated by the fusion reaction in the central area of the sun is transported to the sun's surface and makes its temperature around 5800 K. This energy conversion process in the sun is just as described in the previous paragraph. In other words, the inner shell of the sun shields dangerous EQ to escaping or releasing from the surface and converted their energy to make massive volume of the sun very high temperature. Still, some EQ is directly emitted from the surface in particular from the solar flare. However, fortunately, the earth's atmosphere shields EQ from reaching to the ground surface. The very special circumstance of the earth, such as the distance between the sun and the earth, existence of atmosphere and water and so on, allows the existence of living beings and human beings as well. The earth is a special/unique planet or only one planet in the universe. We human beings are benefiting from the radiated energy as EQ from the sun with shielding the dangerous EQ by the atmosphere. There is no wonder that some people believe God did make the earth and living beings.

1.6   Radioactive Materials and Artificial EQ (Radiation) Sources

Figure 1.5 shows EQ sources all of which are RIs existing in nature. Terrestrial radionuclides having extremely long lives existing on/in the earth from its birth are potassium-40 (^{40}K), rubidium-87 (^{87}Rb), samarium-147 (^{147}Sm), lutetium-176 (^{176}Lu), rhenium-187 (^{187}Re), radon-222 (^{222}Rn), radium-226 (^{226}Ra), thorium-232

Fig. 1.5 Radiation sources in nature

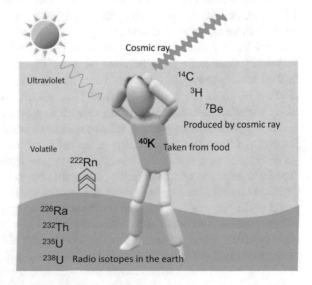

(^{232}Th), uranium-235 (^{235}U) and uranium-238 (^{238}U). ^{222}Rn is present in the atmosphere because it is volatile. The earth is gradually losing its stored energy at
the birth resulting in the decreasing temperature on a 100 million-year scale. This
decreasing temperature tendency is separated from the recent global warming seen
on a yearly scale. Nevertheless, energy release from these RIs has somewhat delayed
the temperature drop.

In addition to these long-life RIs, some EQ coming from outer space (called
cosmic rays) react with oxygen and nitrogen to generate RIs, such as tritium (^{3}H,
or T), beryllium-7 (^{7}Be), and carbon-14 (^{14}C). However, their existing amounts in
nature are kept nearly constant owing to the balance between the generation and
disintegration (decay). The air dose rate in nature shown in Table 1.4 is mostly due
to all these RIs existing in nature and generated by cosmic rays.

Additional air dose rate is given by RIs produced by nuclear bomb tests after
World War II and distributed worldwide. Figure 1.6 shows the time traces of ^{3}H
and ^{14}C concentrations in nature after the second world war. After the significant
increase of the both isotopes, they have been decaying according to their respective
half-lives of 13.6 y and 5730 y owing to the partial nuclear test ban treaty (PTBT)
came into effect in 1963. Nowadays, because of the shorter half-life of ^{3}H, its amount
has returned almost to a pre-war level. However, cesium-137 (^{137}Cs) and strontium-
90 (^{90}Sr) also generated by the nuclear tests are keeping easily detecting levels. In
comparison with the air dose rate in 1963, the present air dose rate is more than 10
times less, showing the annual air dose rate of less than 1 mSv in most places on the
earth. It is unclear how the higher air dose in '60th has influenced the appearance of
cancers. It is said that the effects of such air dose would be far less than those caused
by tobacco, pesticides, stresses, and others.

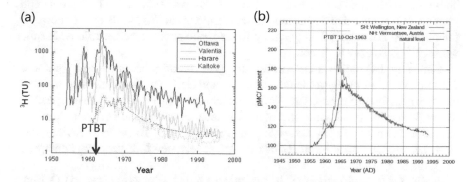

Fig. 1.6 Change of natural radiation levels of **a** Tritium (^{3}H) in Canada and **b** 14-Carbon (^{14}C)
in New Zealand after World War II. After PTBP (Partial Test Ban Treaty) in 1965 both have been
decaying following their half-life. **a** After IAEA/WMO (2006). Open on web site of the Global
Network of Isotopes in Precipitation (GNIP) and Isotope Hydrology Information SIAEA (2006). W.
G. Mock, Environmental Isotopes in the Hydrological Cycle Principles and Applications, IAEA,
2000 (https://www.iaea.org/services/networks/gnip). **b** Source, http://en.wikipedia.org/wiki/File:
Radiocarbon_bomb_spike.svg, open for public

(a)   **(b)**

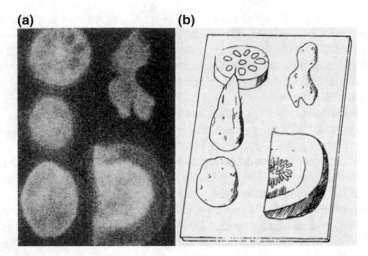

Fig. 1.7 **a** Profiles of ^{40}K included in root vegetables (yellow areas correspond to high ^{40}K density) determined by an imaging plate and **b** their sketches

Furthermore, potassium present in nature contains an RI of ^{40}K. The isotopic ratio in natural K is 93.26% and 6.73% for its stable isotopes ^{39}K and ^{41}K, respectively, while only 0.012% is the radioactive isotope of ^{40}K. Figure 1.7 shows examples of ^{40}K distributions in various root vegetables measured by an imaging plate. One can see that the seeds of a pumpkin include high ^{40}K. A banana is also well known to include significant amount of K so as ^{40}K. One banana eaten gives the absorbed dose equivalent to 0.1 μSv. The presence of ^{40}K is easily detected even in seaweeds that contain a lot of K and can be imaged like those shown in Fig. 1.7. The figures might give the impression that foods contain large amount of ^{40}K. In daily life, however, the absorbed dose equivalent given by foods is too small to be hazardous. In other words, EQ is so easy to detect.

Generally, EQ includes three different types, α- and β-particles, and γ-photons. The α-particles are energetic helium ions, He^{+} and He^{2+}. The β-particle is an energetic electron negatively charged. There is a positively charged electron called as a positron (e^{+} or β^{+} particle) and some specified RIs decay emitting the positron. Any ions accelerated to high velocity or carrying high energy is EQ. Therefore, a device called an accelerator in which electrons or ions are accelerated to have a large energy, is the artificial EQ source.

Again, the term of "exposure" is used when a living being is exposed to EQ. More precisely, all or part of EQ energy is deposited to or absorbed in. Taking the differences in types of EQ the absorbed dose (energy) (Gy) is converted to absorbed dose equivalent (Sv) with the radiation weighting factor (W_R), which is further converted to effective dose with consideration of different characteristics of tissues or organs in a human body with the tissue weighting factor (W_T) as given in Tables 2.1 and 2.2 in Chap. 2, respectively. As already mentioned, when a substance is exposed to

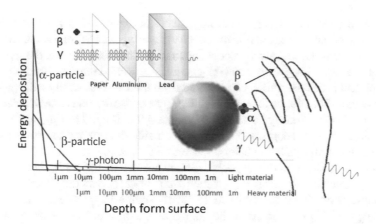

Fig. 1.8 Depth profiles of deposited energy for α- and β-particles, and γ-photon injected from the surface of a substance

EQ, irrespective of the substance is a living being or not, and absorbs 1 J of energy in 1 kg, the absorbed dose is defined as 1 Gy.

The effect of EQ exposure is different depending on the character of the substance-exposed and absorbed or deposited energy is not uniform in depth of the substance. Figure 1.8 shows how three types of EQ, α and β-particles and γ-photons, are penetrating the substance and depositing their energy in depth. In the figure, the area surrounded by energy deposition and the horizontal axis, or integration of the energy deposition rate with depth, corresponds to the total absorbed energy in the substance.

Note that the depth of the horizontal axis is on a logarithmic scale. The figure clearly shows that the penetrating depths are quite different among the three types of EQ. The energy absorption is also different depending on the mass of the substances, i.e., heavier the mass, shallower is the penetration depth, or larger the absorbed (deposited) energy in unit mass or volume. This will be detailed in Chaps. 3 and 4.

1.7 Summary

The term "radiation" is used to represent several physical phenomena, emission of EQ (energetic quanta) including quantum particles and photon form sources, exposure of a substance to EQ, and absorption of EQ energy or energy deposition in the substance. Generally, EQ carries high energy in the order of 10^6 eV/quantum. Any substances exposed to EQ are given (absorb) all or part of the energy of EQ carrying. Physically, the EQ exposure of a substance means that the substance is exposed to power $(W\,m^{-2})$ and absorbs part of or all of the power within unit time as absorbed energy or absorbed dose $(J\,kg^{-1} = Gy\,s^{-1}, \text{ or } J\,m^{-3})$. The absorbed or deposited energy is converted through various physical and chemical processes and finally turned into the heat or temperature rise of the substance. In the energy conversion processes, there is a particular energy range where the biological effects cause cell death, chromosomal abnormalities, canceration, and so on. Since various types of EQ exist in the universe, such energy conversion is always occurring in somewhere. In stars like the sun, fusion reactions generate huge energy and the energy is converted through various physical and chemical processes and finally emitted from their surface as EQ with very wide energy range. From the surface of the sun, the energy emission is like the blackbody radiation at 5800 K mostly consisting of visible light. The EQ having higher energy or dangerous EQ are suppressed by shielding inside the sun and the atmosphere of the earth, which makes living beings survive on the earth.

Exposure to EQ is nothing special. On the surface of the earth, there is always some EQ sources. However, EQ emitted from them is generally too weak to be hazardous. There is no need to fear EQ if its source is specified. Once the type and energy of EQ emitted from the source or existing in the surroundings were determined with a suitable detection technique, we could safely handle it, or even we could convert their energy into useful ones. Nevertheless, the energy conversion processes are not easy to understand. Depending on the type of EQ exposed and on the character of exposed substances (tissues, organs, part of a body), the appearance of the effect of the exposure is different. At lower absorption dose, the appearance of the effect of the EQ exposure is stochastic, probabilistic, or significantly scattered. These uncertainties are very likely the cause to give the scarcity of EQ. However, the EQ exposure can be avoided, and EQ is controllable no need to be "afraid". The main purpose of this book is to explain these points.

Chapter 2
Radiation (Energetic Quanta: EQ)

Abstract This chapter is devoted to clarifying (1) what "radiation" is, (2) what "exposure to radiation" means, and (3) what kind of effects "the exposure to radiation" gives to substances. The term "radiation" is used to represent various physical phenomena and hence it seems to be used with confusion. Its original meaning is emission of electromagnetic waves mostly as visible light from a high-temperature substance. After finding X-rays, the emission of high-energy electromagnetic waves is also referred to as radiation. Then radioisotopes (RIs) were found as a radiation source, which emits radiation, and sometimes the radiation source is also called radiation. Afterward radiation is well known to be hazardous such that the exposure of living beings to the radiation causes damage or certain unwanted effects.

Keywords Absorbed dose · Dose equivalent · Effective dose · EQ (Energetic Quanta) · Exposure · Radiation · Shielding

2.1 Introduction

The term of "*radiation*" is used to represent various phenomena and hence it seems to be used with confusion. The original meaning is emission of electromagnetic waves mostly as visible light from a high-temperature substance. After finding of X-rays, the emission of high-energy electromagnetic waves is also referred as *radiation*. Then radioisotopes (RIs) were found as a *radiation* source, which emits *radiation,* and sometimes the radiation source is also called *radiation.* Afterward, *radiation* is well known to be hazardous such that the exposure of living beings to the radiation causes damage or certain unwanted effects. The confusion appears also in discussion of the radiation on its strength or intensity and energy. The purpose of this chapter is to clarify (1) what "*radiation*" is, (2) what "exposure to *radiation*" means, and (3) what kind of effects "the exposure to *radiation*" gives to substances (inorganic and organic materials and living beings).

As already introduced, the radiation considered in this book is consisting of Energetic Quanta (**EQ**; including both quantum particles and quantized photons). EQ includes high-energy quantum particles so-called elementary particles like proton,

© Kyushu University Press 2022
T. Tanabe, *Radiation: An Energy Carrier*,
https://doi.org/10.1007/978-981-19-1957-2_2

neutron, α-particle, heavy-ion, electron, positron, and so on, and quantized electro-magnetic waves (photons) like γ-ray, X-ray, ultraviolet light, and so on. Hence to characterize an EQ source, intensity or number of EQ emitted from the source in unit time which is referred to as "radioactivity", and type and energy of each quantum should be determined. On the other hand, for the exposed side, the number of EQ exposed in unit time and in unit area and the type and energy of each quantum should be determined. Furthermore, for discussing the effects of exposure appearing in a substance, it is critically important to know how much energy of EQ is deposited on the substance or the substance absorbs because the substance exposed to EQ does not necessarily absorb all energy of EQ. Moreover, for different types of EQ, the effects of their exposure are not necessarily the same even if the absorbed energy is the same. In other words, the energy absorption or deposition processes in a substance exposed to EQ are totally different depending on the type of EQ and the characteristics of the substance.

Thus, to characterize the source of EQ, the type of EQ emitted, and the intensity and energy of EQ should be measured. The major EQ source on the earth is radioactive isotopes or radioisotopes (RIs) whose nuclei disintegrate emitting specified EQ. If the EQ source is an RI, the intensity of EQ, i.e., the number of EQ emitted from the RI in unit time is corresponding to the number of nuclear disintegration or decay of the RI (dps, dpm, or dph, depending on the time unit of second, minute or hour, respectively) and dps in unit weight is represented with the SI unit of **Becquerel** (Bq/kg). Although the term "radioactivity" is generally used to represent the intensity of EQ emitted from RI, it is also used to express the intensity qualitatively, like strong or weak radioactivity. In this book, "the intensity of EQ emitted from a source (EQ intensity)" is exclusively used instead of the "radioactivity" of RI or the EQ source.

The source of EQ is not limited to RI in nature, but various artificial EQ sources are produced and some arrive from the universe. The intensity of EQ emitted from these sources is measured by a radiation detector and is represented with using detected counts in unit time (cps, cpm, or cph, i.e., counts per second, minute and hour, respectively). Although the detected counts are often referred to as the radioactivity of the sources or air, they do not directly correspond to the intensity (radioactivity) of EQ source given in Bq. It requires some calibration considering geometrical relations between a detector and an EQ source to determine its radioactivity.

From the viewpoint of EQ exposure or a substance exposed to EQ, more important is how much energy is given or deposited by EQ, or absorbed in the substance. Absorbed or deposited energy in the substance is given with the unit of Joule (J) in its unit mass or volume, i.e., J kg^{-1} or J m^{-3}. Since the manner of energy absorption is quite dependent on the mass of the substance, as described later, the unit of J kg^{-1} is employed as absorbed or deposited energy and represented with the unit of **Gray** (Gy) (1 Gy $= 1$ J kg^{-1}) and the absorbed energy is called as "**absorbed dose**". The word of "**dose**" would have been introduced at the beginning of investigation of the effects of EQ exposure to represent radiation as something countable because the nature of the radiation was not well known. Nowadays the radiation is well defined and it is well known that the dose is equivalent to the amount of energy. The author would like to recommend to use "**absorbed energy**" instead of the absorbed dose.

Since the manner of energy absorption (deposition) is different depending on the type of EQ, introducing a factor called "**radiation weighting factor (W_R)**", the difference of the type of EQ is normalized and represented with the unit of **Siebert** (Sv) (1 Sv = $W_R \times$ 1 Gy) and referred as "**absorbed dose equivalent**", as given in the next section. Furthermore, in case of the exposure of a human body, energy absorption rate varies depending on the part of the body organs or tissues. Hence an additional factor, called "**tissue weighting factor (W_T)**", is introduced to normalize the difference with conversion of absorbed dose equivalent to "**effective dose**" (see Sect. 2.4).

In this chapter, after introducing sources of EQ, details of absorbed dose, dose rate, dose equivalent, and effective dose are explained. Since EQ sources are always distributed in atmosphere or surroundings as natural radiation, EQ exposure could be caused by air from which energy deposition is referred to as "air dose" and "air dose equivalent".

2.2 Radiation is Consisting of Energetic Quanta (EQ)

Concerning "radiation", following three points should be noted in this book.

1. "Radiation" is consisting of high-energy quantum particles and quantized light (photons) referred to as energetic quanta (EQ).
2. A radioactive material emitting EQ includes RIs.
3. Instead of "radioactivity", the term "intensity of emitted EQ or simply EQ intensity" of the radioactive material is used.

There are various EQ sources, like RIs which contain extra energy in their nuclei and emit the extra energy as EQ, fission reactions in nuclear reactors, fusion reactions in the sun, and so on. Cosmic rays are also EQ from the universe. Since "radiation" and "radioactivity" are not clearly defined, as described above, these two terms are basically not used hereafter in this book, instead, EQ and intensity of EQ emitted from EQ sources are used.

EQ are separated into two kinds of quanta, i.e., photons and quantum particles. The former are electromagnetic waves without mass and the latter with mass including various elementary particles (muon, pion, mesons, neutrinos, etc.), electrons (β–particles), positrons (β^+–particles; positively charged electrons), protons, neutrons, α–particles, various kind of ions (usually positive ions, and some negative ions which are talked about in health devices, etc.), or fission products. EQ is also separated into those making ionization of molecules and atoms (ionizing radiation), and the others having no ability to ionize (non-ionizing radiation). Since most of quantum particles have a charge, they are ionizing radiation (EQ), or charged quanta. Since neutrons and neutrinos have no charge, they are categorized as the non-ionizing radiation. However, neutrons are very dangerous because they do not receive a Coulomb repulsion to and from the nucleus, and they can easily reach the nucleus to give energy causing a nuclear reaction. Nuclear reactors use the fission reaction of uranium (^{235}U) with neutrons producing fission products and releasing fission energy. Neutron bombs are also developed to enhance neutron emission and killing ability. The neutrons are always generated in fission and fusion reactions in nuclear reactors. However, they are usually shielded so that people outside the reactor are not exposed to them. In addition, a single neutron is not stable and decays into a proton and an electron with half-life of 7 min.

Energetic photons or electromagnetic waves collide with electrons of constituent atoms of a substance to be in excited or ionized states which are referred to as the Compton scattering.

The energy carried by photons (electromagnetic waves), $E,$ is given as

$$E = h\nu = h/\lambda \tag{2.1}$$

where ν and λ are the frequency and the wavelength of the photon and h the Plank constant. As shown in Table 1.1 in Chap. 1, photons having energy of more than around 10^5 eV are referred to as γ–rays, around 10^5–10^2 eV, X-rays, around 1000–100 eV soft X-rays, 50–10 eV, and ultraviolet light around 10–5 eV. Since photons with energy above around 10 eV are able to excite or ionize the constituent elements of a substance, they are referred to as the ionizing radiation. Photons having the energy of less than around 5 eV do not ionize the elements and are referred to as the non-ionizing radiations. They are visible light, around 5–0.01 eV, infrared light, less than around 0.01 eV, and far-infrared light. Those having less energy are electromagnetic waves. From ultra-high-frequency band (GHz band), the shortwave band (MHz band), the mid-wave band (kHz band), and the long-wave band (Hz band). It is well known that the FM band is used for the television and the kHz band (called the medium wave) as radio waves. It may be difficult to recognize that these low-energy electromagnetic waves are physically the same kind with γ–and X-rays, i.e., only different in wavelength or frequency.

A photon with energy above around 1 meV can be detected, while electromagnetic waves with lower energy than infrared light are difficult to detect as isolated photons. When large number of photons are exposed to a substance, they are detected as its temperature rise. Intense exposure to high-frequency electromagnetic waves, such as those used in microwave ovens using 2.45 GHz, shows a significant impact on the human body, such as burns, if shielding (described below) is not well done. In electric heaters commercially available, the conversion of electricity to far-infrared light is effectively done.

Charged quanta lose their energy (slow down) due to Coulomb interactions or collisions with electrons or nucleus of constituent elements of a substance and excite or ionize the constituent elements. That is the reason why they are referred also to as the ionizing radiation. In other words, they give (deposit) the energy to the substance through Coulomb interactions. When their energy deposition rate is large, they deposit all of their energy in the material after moving only short distance and stop, while the energy deposition rate is small, they penetrate deep. In the following, a little detail of the energy deposition processes is described.

2.3 Sources of EQ and Their Intensity (Radioactivity)

2.3.1 Sources

Except for cosmic rays, EQ is always released from emission sources on the earth. Although the details of the sources are discussed in Chap. 3, briefly introduction is given here. An accelerator is an artificial EQ source that gives energy to charged particles by accelerating with electromagnetic field, and they are also indispensable for the study of natural science in particular high-energy physics. Whether useful or not, EQ sources are separated into natural or artificial ones. Nevertheless, there is no difference between the two when their type and energy are the same. Some people believe that artificial one is different from the natural one, but that is a complete misunderstanding. The major artificial EQ sources are accelerators, RIs made of nuclear reactors, and even nuclear bombs. The so-called death ash or nuclear fallout is dust or microparticles containing various radioactive fission products (FPs or RI) made from nuclear explosions (nuclear reactions proceed explosively).

Since the details of RIs are described in Chap. 3, here are briefly introduced. An RI is an element that has extra energy in its nucleus and the excess energy is released as EQ and converted into a stable nucleus (stable isotope) which is referred as disintegration or decay of the nucleus. The emitted EQ at the disintegration of an RI are either α–and β–particles or γ–photons, referred to as α-, β-, or γ–decays, respectively. At α–decay, the nucleus emits $^4_2$He. Accordingly, the new element has atomic number (A) and mass number (M) smaller by 2 and 4 ($^{M-4}_{A-2}Z'$) than those of the original element ($^M_A Z$ with Z, the name of the original element). In β–decay there are two types, one emits an electron (e^-) and the other a positively charged electron named a positron and referred β^+-decay. After the β–decay, the atomic number increases by 1, while the β^+-decay decreases by 1. In γ–decay, there appear no changes in both the atomic number and mass number, because γ–photon is an electromagnetic wave with no charge. The decay stops when an RI emits all excess energy. The energy release rate decreases exponentially with a specified half-life for each RI.

There are various RIs of cosmic origin. Different from planets in cosmos, all stars emit energy, i.e., they are the sources of EQ. In the universes far from the earth, space explosions occasionally occur and huge energy is released in all directions and dispersed. Accordingly, various kinds of EQ are existing in the universe. Neutrinos are also of cosmic origin, for which detection a particular device is set in several places on the earth, one example is Super Kamiokande in the Kamioka Mine in Gifu Prefecture in Japan [1]. Tritium, a radioactive isotope of hydrogen (^{3}H or T) has two origins. One is produced by cosmic rays with their nuclear reaction with nitrogen and oxygen, and the other is artificially generated mostly by nuclear reactions in nuclear bomb tests and some by nuclear reactors. Tritium is converted (nuclear decay) into 3-helium (^{3}He), a stable isotope, by the β–decay with half-life of 13. 6 years. The natural abundance of T is nearly constant, because, T production by cosmic rays and nuclear decay is balanced. However, owing to the nuclear bomb tests from the 1950s to the 1970s, it had been significantly increased from the natural level as shown in Fig. 1.6 in Chap. 1. After PTBP (Partial Test Ban Treaty) in 1965, the T production stopped and its abundance has been decreasing according to its half-life and now mostly returned to the natural level. Still burning of fossil fuels in thermal power plants around the world, is releasing a detectable amount of T as HTO, because the fossil fuels contain trace amounts of natural tritium. Nuclear power plants also generate some T. Hence, the present abundance of T in nature is a little higher than that of a hundred years ago without artificial release.

Since the sun continues to emit energy through fusion reactions as a star, various EQ are emitted from its surface. However, owing to long distance between the sun and the earth (the light takes about 8 min to reach the earth), and the existence of the atmosphere of the earth, higher energy parts of EQ which are dangerous to living beings are shielded. Accordingly, solar energy, i.e., the energy is given by the sun consists of a little ultraviolet light, mostly visible light and some infrared light. Consequently, human beings get benefit from the solar energy. Since the moon does not have atmosphere, high-energy part of EQ directly impinges on its surface. The role of spacesuit for astronauts is not only to keep air inside but also to protect the EQ exposure. In recent years, the space development draws much attention. Nevertheless, protection or shielding of EQ (cosmic rays) in the space is one of the key factors preventing human beings from entering the space.

Since inter-continental flights travel over an altitude of about 10,000 m, passengers cannot avoid the exposure to the cosmic rays. Although the body of the aircraft shields some, they are still exposed to about 0.2 mSv in one round trip between Tokyo and New York, for example.

2.3.2   Radio Isotope (RI)

As is well known, Madam Curie noticed the release of radiation from natural uranium because it contains a radioactive isotope ^{238}U making α–decay to be ^{234}Th emitting α–particle (^{4}He). Since the emission of EQ from an RI is due to disintegration of its nucleus (nuclear decay), it occurs randomly. Figure 6.2 in Chap. 6 shows a time sequence of EQ emission by the nuclear decay. The EQ emission appears randomly. Nevertheless, the nuclear decay obeys a certain rule (with a specified half-life).

Suppose an RI contains N radioactive nuclei, the number of nuclear disintegrations per unit time is known to be proportional to presently existing ones.

$$dN/_{dt} = -\lambda N \tag{2.2}$$

where λ is a proportional constant referred as the decay constant, and every RI has its own value. According to (2.2), for an RI initially containing N_0 radioactive nuclei, the number of remaining nuclei at time t is given by,

$$N(t) = N_0 \exp(-\lambda t) \tag{2.3}$$

The time period (τ) until the initially contained radioactive nuclei become a half is.

$$\tau = \ln 2/\lambda. \tag{2.4}$$

Thus, the time period is a constant regardless of the initial number of N_0, and is called the half-life of the RI. The number of disintegrations (decaying nuclei) per unit time decreases following Eq. (2.3). Hence any RI even if it has extremely long half-life, decreases with time. However, the number of disintegrations in unit time is nearly constant on daily scale for RIs having a very long half-life of more than 100 years. "Radioactivity" is used to represent not only the number of nuclei disintegrating in unit time for a particular RI but also the number of EQ detected in unit time whatever the sources are.

As mentioned earlier, the number of disintegrating nuclei per second or per minute (described as dps or dpm, respectively) is defined as the intensity of EQ emitted from or radioactivity of RI or an EQ source, and represented with a unit called Becquerel (Bq) as 1 Bq = 1 dps. This book uses definitively the intensity (instead of the radioactivity) of the EQ source and makes comparison in unit volume (Bq/m^3) or unit mass (Bq/kg).

It should be noted here that the EQ intensity given by Bq does not contain any information on the kind and energy of EQ. Even if two different RIs have the same intensity (radioactivity), i.e., the same Bq numbers, the effects of EQ exposure to a substance, including a human body, are not necessarily the same. Because the types and energy of EQ emitted are different for each RI. For example, T emits β–electrons with an average energy of only 5 keV, while iodine (^{131}I) and cesium (^{137}Cs) in Fig. 3.4 in Chap. 3 emit much higher energy β–electrons. Therefore, the latter two are more dangerous to a human body compared to the former. Furthermore, ^{131}I and ^{137}Cs emit simultaneously γ–photons which are more dangerous. As such, different types and energy of EQ are emitted from each RI with its EQ emission rate decreasing with their half-lives.

2.3.3 Geometry of EQ Sources (Point, Volumetric, Planar and Spatial Sources)

Although the EQ exposure is evaluated as the incident number and absorbed (deposited) energy of EQ in unit volume or weight of a substance, as described in detail in Chap. 6. It is important to know how EQ is emitted from a source, i.e., what types of EQ are emitted and how intense their emitted number, and how high their energies are. In this section is described how the intensity of EQ changes with the geometrical structure or shape of EQ sources.

2.3.3.1   Point and Volumetric Sources

If an EQ source is a solid or liquid, it is called a volumetric source because it emits EQ in all directions, and the number of emitted EQ, or the intensity of EQ is represented with the number of Bq per unit volume (Bqm^{-3}), or unit mass ($Bq\ kg^{-1}$) of the source. When the source is small or placed far, it is treated as a point source.

When the source has a large volume, the emitted number of EQ from the source might not be the same with the total number of disintegrating nuclei in the source, and energy of EQ emitted from the source might not be the same with that generated inside the source. Because EQ emitted in the inside of the source loses their energy before coming out from its surface, as shielding effects. The sun is a typical example. The sources emitting α– or β–particles are other examples. These particles easily lose their energy during their passing through a source material and cannot escape from deep inside, which is referred to as self-shielding. For such cases, only EQ emitted from the near-surface is concerned and it would be treated as a surface source as described in the next section, although the source itself is volumetric.

On the contrary, an EQ source taken inside of a substance (or into a body) is quite hard to characterize (measure) the intensity and energy of EQ emitted owing to the shielding effects. For such cases, the EQ intensity is measured by sampling some volume of the substance and grinding it to be liquid-like so that the intensity is represented with $Bq\ kg^{-1}$. For the internal exposure, it is quite hard to determine the actually absorbed dose equivalent or effective dose given by the source, because RIs in the source are included volumetrically.

2.3.3.2   Planar Source

When EQ sources adhere to some surface, or they are emitting α– or β–particles, the number of EQ emitted per unit area of the surface, i.e., the flux of EQ in unit time passing through unit area is concerned ($Bq\ m^{-2}\ t^{-1}$). Since EQ given by accelerators and/or nuclear reactors tend to be emitted unidirectionally, they are also considered a planar source and the intensity of emitted EQ is handled with EQ flux in unit area and unit time ($Bq\ m^{-2}\ t^{-1}$).

2.3.3.3   Spatial Source

In air, fine particles containing radioactive isotopes are floating as dust. They contain ^{60}Co and ^{137}Cs, which have been distributed in past nuclear tests. In addition, ^{14}C and T generated by cosmic rays are contained in the air as carbon dioxide or tritiated water vapor. It also contains ^{222}Rn, which is produced by the nuclear decay of natural uranium and thorium. Since they are homogeneously distributed in the air, the intensity of EQ in the air is often presented as air dose rate ($Gy\ s^{-1}$) or air dose equivalent ($Sv\ s^{-1}$). It is well known that the air dose rate near the ground becomes high during rainfall because the dust float down (see Fig. 2.1). Since ^{222}Rn is contained in lime

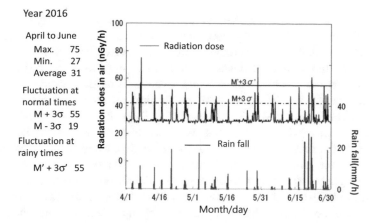

Fig. 2.1 Changes in air dose rate and rainfall in the vicinity of Genkai Nuclear Power Plant, Saga Prefecture, Japan (April-June 2016). Announced at the Saga Prefectural Conference on Environmental Radioactivity Technology. [Prepared by Nuclear Safety Division, Saga Prefectural Government]

and soil of the ingredient of concrete, the air dose rate in a concrete room is higher than in wooden buildings or the open air.

2.3.3.4   Natural Radiation (EQ Exposure in Nature)

In the previous section, are introduced the sources of EQ and the unit to represent their radioactivity or intensity as Bq. On the other hand, for consideration of the effects of EQ exposure, more important is how much energy is deposited or absorbed in a substance exposed to EQ, which is given with a unit called absorbed dose, gray $(Gy : 1\,Gy = 1\,J\,kg^{-1})$, or absorbed dose equivalent, Siebert (Sv), which are discussed in more detail in this section.

The air dose indicates how much energy is absorbed in a human body exposed to EQ emitted from radioactive isotopes in air and on the ground. As depicted in Fig. 1.8 in Chap. 1, if the types of EQ in air are not specified, the intensity of EQ is given by the number of EQ detected as cps, cpm, and so on. However, in the normal atmosphere α–particles and β–particles easily lose their energy and stop within a few cm at most (only a few cm they can proceed). Therefore, as described later in this section, the air dose or air dose rate is estimated for a human being exposed to γ–photons at the specified location. A simple instrument to measure the air dose named a dosimeter indicates thus estimated absorbed dose or dose equivalent in Gy or Sv.

Figure 2.1 shows the daily change of air dose rate measured near Genkai nuclear power plant in Saga prefecture, Japan, from January to March in 2017 [2]. During this time the plant was out of operation. Measurements were performed with an ionization chamber set at 1.5 m above the ground. Although the average value of the measured air dose rate was 30 nGy h^{-1} (where nGy is 10^{-9} Gy), they varied day by day, even time to time, sometimes showing more than two times larger than the average, about 60 nGy h^{-1}. Similar changes and the average value were observed under the plant operation, i.e., the air dose rate was raging from 20 to 60 nGy h^{-1} irrespective of the plant operation. The main cause of the deviation is rainfall. Most of time when the air dose was high, it was raining. When the air dose rate was significantly high, there should be an artificial release somewhere. Thus, the release from the nuclear accidents at Chernobyl and Fukushima was detected as the increase in the air dose.

Normally, the natural air dose is due to radioactive isotopes contained in floats in air, and on the ground surface. Therefore, rains fall down the floats and deposit them on the ground to increase the air dose rate near the ground surface in wide area as shown in Fig. 2.1. On the other hand, if radioactive materials (EQ sources) were dispersed into the air, as appeared in the Fukushima nuclear accident, the increase of the air dose rate should be localized near the source or spread along the direction of wind. Furthermore, they fell to the ground surface. Therefore, the increase of the air dose rate was not uniform showing higher and lower air dose areas. Accordingly, the measurement of the air dose rate by a detector shows directionality, i.e., the air dose rate varies greatly depending on the direction of the detector and the distance from the ground. Applying this, one can determine where EQ sources are deposited on the ground surface.

2.4 Energy Absorption (Deposition) Given by EQ Exposure

As mentioned earlier, there are various types of EQ, and the manner of energy absorption or deposition in a substance exposed to EQ varies with not only the type of EQ but also characters of the substance, such as chemical composition, mass, and temperature.

In principle, EQ entered in a substance loses some or all of its energy through collisions with electrons and atoms in the substance. Each collision, an energetic quantum loses its energy by $\Delta\varepsilon$,

$$\varepsilon_{in} = \varepsilon_{out} + \Delta\varepsilon \tag{2.5}$$

where ε_{in} and ε_{out} are the energy before and after the collision. As introduced in Fig. 1.1 in Chap. 1, for a quantum particle, its energy is carried as velocity. Hence the energy loss is represented as the velocity change from v_{in} to v_{out},

$$\frac{1}{2}m \cdot v_{in}^2 = \frac{1}{2}m \cdot v_{out}^2 + \Delta\varepsilon \tag{2.6}$$

For a quantized photon, the energy loss appears as the change of frequency (v) or wavelength (λ),

$$h \cdot \nu_{in} = h \cdot \nu_{out} + \Delta \varepsilon \qquad (2.7)$$

or

$$h/\lambda_{in} = h/\lambda_{out} + \Delta \varepsilon \qquad (2.8)$$

Since the energy loss occurs as quasi-continuum in substances, the energy loss depends on the depth. Using more general expressions as given Table 1.3 in Chap. 1, the energy loss for a particle moving from d_1 to d_2 in depth is given as.

$$\Delta \varepsilon = \int_{d_1}^{d_2} \frac{\partial \varepsilon}{\partial x} dx, \qquad (2.9)$$

where $\frac{\partial \varepsilon}{\partial x}$ is the stopping power of the substance for an energetic quantum. When collisions with electrons are dominated, $\frac{\partial \varepsilon}{\partial x}$ is constant and such energy loss process is called **linear energy transfer** (LET) regime as discussed in Sect. 4.2 in Chap. 4. Since the lost energy is given to the substance, the energy loss of the energetic quantum is the same as the absorbed or deposited energy in the substance.

Although the term the "absorbed or deposited energy" might not be familiar, the effects appeared in the substance exposed to EQ are caused by some or all of EQ energy absorbed or deposited in the substance. From the viewpoint of the exposure of living beings, "EQ exposure" means that absorption (deposition) of all or part of the EQ energy referred to as "absorbed dose".

Figure 1.8 in Chap. 1 shows depth profiles of absorbed (deposited) energy for α– and β–particles and γ–photons injected into a material with the same energy. The α–particles deposit all energy within 1 µm because of their heavy mass or strong interaction (large stopping power) with electrons and atoms in the material. The β–particles penetrate a little deeper. Nevertheless, their penetration depth is within a few mm at maximum. Different from these particles, the γ–photons, which are electromagnetic waves, penetrate over 1 m range. The areas surrounding the depth profiles of each EQ and the baseline or integration of deposited energy with the depth in Fig. 1.8 in Chap. 1 represents the total absorbed (deposited) energy in the material. Therefore, when a human body is exposed to α– or β–particles, the effects appear mostly in skin and tissues in little deeper regions as the possible causes of "skin cancer" and "radiation burns", and less effects in tissues or organs in the body compared to the exposure to the γ–photons. Nevertheless, as explained later, when an EQ source is taken into a body, which is referred to as internal exposure, the effects become appreciable. The energy absorption (deposition) caused by the exposure of γ–photons is rather uniform throughout the body.

If a material subjected to EQ is heavy, for example, lead (Pb), the penetrating depths of EQ are shorter than that for lighter materials as indicated in the horizontal scale in Fig. 1.8 in Chap. 1. In other words, installation of heavy material in front of a substance (a human body) can suppress energy absorption (deposition) in the substance. This principle referred as a **shielding effect** is employed for the protection of EQ exposure and a heavy material (Pb in most cases) is used as the shield.

2.5 Energy Absorption in Living Beings Exposed to EQ

Again, EQ exposure to living beings means that the living beings is subjected to EQ and all or part of the EQ energy is absorbed or deposited in their body. Then the absorbed energy is converted to absorbed dose equivalent using the radiation weighting factor (W_R) given in Table 2.1 [3]. Therefore, if the absorbed dose equivalent were the same, the effects of the exposure of different types or different energy (profiles) of EQ should be the same. However, the manner of energy absorption or energy deposition processes is different with the energy distribution of EQ and the types of EQ. The main difference is in energy deposition profiles in depth. The absorbed or deposited energy profiles in depths are different depending on the energy distribution of EQ even for the single type of EQ. The difference in the absorbed energy profiles in depth is much more significant among α–particles, β–particles, and γ–photon penetrating within μm, mm, and over m, respectively as shown in Fig. 1.8 in Chap. 1. These differences would be appreciable between the internal and the external exposures of α–particles and β–particles. Hence both are explained independently.

Table 2.1 Radiation weighting factors (W_R) for conversion of absorbed dose rate (Gy s^{-1}) to absorbed dose equivalent rate (Sv s^{-1}) according to ICRP 103 (ICRP 2007) [3]

Kind of EQ	Weighting factor (W_R)
Photons (X-ray, γ–photon)	1
Electron (β-particle), muon	1
Neutron <10 keV	5
Neutron 10–100 keV	10
Neutron 100–2000 keV	20
Neutron 2000–20,000 keV	10
Neutron >20,000 keV	5
Proton (Except recoil proton) >20,000 keV	5
α–particle	20
Fission product	20
Heavy nucleus	20

2.5.1 External Exposure

Generally speaking, EQ exposure (radiation exposure) is the exposure of living beings to EQ, and as a result, the energy of EQ is absorbed (deposited) in skins, tissues, and organs in a human body. In respect of energy absorption, there is no difference between the EQ exposure and exposure to infrared light radiated from a heater. When energy is locally absorbed on the surface of the skin, it becomes burns or necrosis and there is no fundamental difference between radiation burns and infrared burns, which are the result of the death of cells in the skin caused by the deposited energy. However, the energy of each photon of visible light and infrared light is so small that without their intensive exposure the burn does not occur. On the other hand, the energy of each energetic quantum is so large that it penetrates the body and deposit its all energy in quite tiny area and even one energetic quantum could cause the death of a cell. Of course, the exposure to one energetic quantum does not appear as any symptom, but a countable number of EQ exposure in very localized area could give. As shown in Fig. 1.8 in Chap. 1, during the penetration into the body, EQ loses energy and deposit energy or the body absorbs the energy. Hence the absorbed energy in the body changes in depth. Since the penetration depth of $\alpha-$ and $\beta-$particles are shallow, their exposure from outside of a body does not give appreciable effects inside the body, while $\gamma-$rays exposure does. In conclusion, for the external exposure, concerned are $\gamma-$rays and very high-energy $\beta-$particles.

2.5.2 Internal Exposure

Usually, EQ exposure occurs as the external exposure as discussed above. However, when an EQ source is injected into a human body as a food or swallowed as a liquid or gas, and taken into the body tissue, the influence of the EQ exposure appears more appreciably compared to the external exposure. From the viewpoint of energy absorption, there is no difference between the external and internal exposures. However, in the internal exposure, an EQ source locates at tissues or organs in the body and they are directly exposed to EQ, and absorbed energy is much larger than that given by the external exposure.

It should be noted that in the case of the external exposure, EQ sources can be easily removed or taken away. Even if they adhered to the skin surface, they can be washed out by taking a shower, etc., which is referred to as decontamination. On the other hand, when an EQ source or an RI is injected into the body and remains in its tissues or organs, it is difficult to remove, and the internal exposure continues. A well-known way to remove the EQ source is drinking a large amount of water, or taking a stable isotope and replacing the RI in EQ source with it isotopically or chemically. The removal is assisted by metabolism as biological discharge. In the biological discharge, the radioactivity of the EQ source decreases following the decay equation given as Eq. (2.1) and shows exponential decay with a specified half-life referred to as a biological half-life. Although the physical half-life (disintegration of an RI) cannot be changed, the biological half-life may be reduced. For example, tritium (T) taken into a human body is usually discharged with the biological half-life of about a week. However, drinking large amounts of water enhances the discharge because T as an RI of hydrogen is easily replaced by H in H_2O and exhausted as body fluids like sweat and urine. Drinking beer, in particular, is known to be more effective than the drinking water, owing to its urination effects, and beer is often placed at the exit of tritium research facilities. The removal of EQ sources is revisited in Sect. 2.6 as decontaminations.

2.5.3 Absorbed Dose, Dose Rate, Dose Equivalent, and Effective Dose

As the measure of EQ exposure, physical units of "absorbed dose" and "absorbed dose equivalent", and their rates in unit time are employed. The absorbed dose means absorbed energy in a human body exposed to EQ and it in unit time is the absorbed dose rate. Here discussed are details of absorbed dose and absorbed dose equivalent presented in the units of Gray (Gy) and Sievert (Sv), respectively.

Suppose 100 million (10^8) γ–photons with the energy of 1 MeV (1×10^6 eV = 1.6×10^{-13} J) is injected into a human body weighing 60 kg in one hour and all their energy is absorbed, then the absorbed dose becomes

$$10^8 \times 1.6 \times 10^{-13} (J) \div 60 \, (kg) = 2.7 \times 10^{-7} (Gy). \qquad (2.10)$$

Since the energy absorption processes in the human body exposed to EQ are different depending on the types of EQ and their energy, as well as the character of the body (mass and volume) and temperature, their differences are normalized by employing empirical parameters.

To normalize differences in energy absorption for different types of EQ, a unit referred to as dose equivalent, Sv, is introduced using the radiation weighting factor, W_R, as

$$H = W_R \times D, \qquad (2.11)$$

where, D is the absorbed dose and H, the absorbed dose equivalent, i.e.,

$$W_R \times 1 \, Gy \, = W_R Sv. \qquad (2.12)$$

W_R for different types of EQ are given Table 2.1. In case of simultaneous exposures with different types of EQ, the total exposure dose equivalent (H_T) becomes the summation of the exposure dose ($D_{T,R}$) multiplied by the radiation weighting factor of each type of EQ.

$$H_T = \sum_R W_R \times D_{T,R} \qquad (2.13)$$

As seen in Table 2.1, there is large difference in W_R depending on the type of EQ. W_R for heavier ions is very large because they deposit energy in quite local area with high density. Neutrons easily collide with a nucleus to trigger a nuclear reaction producing radioactive nuclei. The exposure with γ–photons or β–particles (electrons) is selected as a reference to examine the irradiation effects so that their W_R is set to be 1. The exposure to γ–photons and β–particles gives similar effects because single γ–photon collides with an electron to exchange their energy referred to as Compton scattering. Hence the absorbed dose is given in Eq. (2.10), 2.7×10^{-7} Gy, becomes the absorbed dose equivalent of 2.7×10^{-7} Sv or 0.27 μSv for the exposure of γ–photons and β-electrons, 20 times larger for α-particles for example. The different appearances of the exposure among the types of EQ are discussed in more detail in Chap. 3.

In case of the EQ exposure of a human body, the appearance of the exposure effect is different depending on the type of tissues or organs, which is compared to radiation sensitivity of each tissue or organ. Therefore, an additional factor referred to as the tissue weighting factor (W_T) is introduced. The tissue weighting factors for different tissues are given in Table 2.2 [4]. Averaging differences of W_T for each tissue, the effective dose (E) given in Eq. (2.14) is introduced to discuss the exposure effects on a human body.

$$E = \sum_T W_T \times H_T = \sum_T W_T \sum_R W_R \times D_{T,R} \qquad (2.14)$$

The amount Sv or μSv appeared in mass media is the value of this effective dose per unit time mostly in an hour or a minute and sometimes a year. The introduction of the effective dose might not be easy to accept or understand. However, because the EQ exposure in most cases is given by γ–photons and/or β–particles, the effective

Table 2.2 Tissue weighting factors (W_T) according to ICRP 103 (ICRP 2007) [3, 4] to convert absorbed dose equivalent to effective dose

Tissue	Tissue weighting factor (W_T)	ΣW_T
Bone-marrow (red), Colon, Lung, Stomach, Breast, Remaining tissues[a]	0.12	0.72
Gonads	0.08	0.08
Bladder, Esophagus, Liver, Thyroid	0.04	0.16
Bone surface, Brain, Salivary glands, Skin	0.01	0.04
	Total	1.00

[a]Remaining tissues: Adrenals, Extrathoracic region, Gall bladder, Heart, Kidneys, Lymphatic nodes, Muscle, Oral mucosa, Pancreas, Prostate (♂), Small intestine, Spleen, Thymus, Uterus/Cervix (♀)

dose is nearly the same as the absorbed dose equivalent. However, for the internal exposure, the effective dose becomes important.

One should remind of the differences in energy absorption and its rate (equivalent to power), absorbed dose and its rate, dose equivalent, and its rate. Absorbed dose given in Gy does not give any information on how many hours of exposure resulted in the value. The absorbed dose is always given at certain times of exposure with a certain absorbed dose rate, or

$$\text{Total absorbed dose} = \int (\text{Absorbed dose rate}) \times dt \qquad (2.15)$$

The same is also for both absorbed dose equivalent and effective dose. The air dose equivalent is often given with the unit of Sv, which is usually the absorbed dose equivalent in one hour ($Sv\ h^{-1}$) for each person. The air dose equivalent for natural radiation in Japan is about 2400 μSv (= 2.4 mSv) per year, which is equivalent to 200 μSv per month, 7 μSv per day, and approximately 0.3 μSv per hour. In Japan, medical exposure is larger than that given by the natural radiation with 3700 μSv per person per year. As mentioned above, the natural dose equivalent of 0.3 μSv/h is almost equivalent to that given by the exposure given in Eq. (2.10) with 10^8 of 1 MeV γ–photons in one hour. Since the mass of a human body is not heavy enough to deposit all energy of 1 MeV γ–photons, i.e., the significant part of γ–photons pass through the body without fully losing their energy as can be seen in Fig. 1.1 in Chap. 1.

Since the absorbed dose equivalent of 200 μSv corresponds to the absorption of energy of 200 μJ per kg of body weight with the radiation weighting factor of 1, the temperature increase of a person of 60 kg caused by this exposure can be calculated as,

$$200\,(\mu J) \times 60\,(kg) \div 4.2\,(J/cal) \div 60\,(kg) \div 1\,(°C/cal) = 0.000000047\ °C. \qquad (2.16)$$

Here the specific heat of the person is assumed to be 1, same as water. If the exposure dose equivalent was 1000 times higher, some exposure effects should appear. Even so, the total absorbed energy would be still 12 J and the temperature increase would be 0.000047 °C. One can see how small the total deposited energy is given by the EQ exposure.

The following compares the energies given to a human body by the EQ exposure and by sunlight. Solar energy is given by electromagnetic waves of ultraviolet, visible, and infrared light. Compared to γ–photons, the wavelength of the sunlight is very long and cannot penetrate through the skin. Hence most of its energy is deposited on the surface of the body or skin. It is known that the solar energy flux in unit time or solar power in unit area is about 100 W m^{-2}. This power deposition can be converted to the absorbed dose equivalent. Suppose the effective surface area of a person with the weight of 60 kg perpendicular to the sunlight being about 1 m^2, energy given in unit time or deposited power is given by 100 (W) $\div$ 60 (kg) = 1.7 W/kg, which is equivalent to the absorbed dose of 1.7 Gy s^{-1} or the dose equivalent of 1.7 Sv s^{-1}, or 6 kSv h^{-1}. Since significant part of the sunlight is reflected, the absorbed energy would be much less. Even if 90% was reflected, the dose equivalent should be 0.6 kSv h^{-1}. If this dose equivalent was given by γ–photons, appreciable effects (cancers or even death) of the exposure should appear (see Fig. 2.2). This is one of the most important characteristics of the EQ exposure on the appearance of the exposure effects, i.e., in spite of small energy or power given by EQ, the effect of the EQ exposure easily appears.

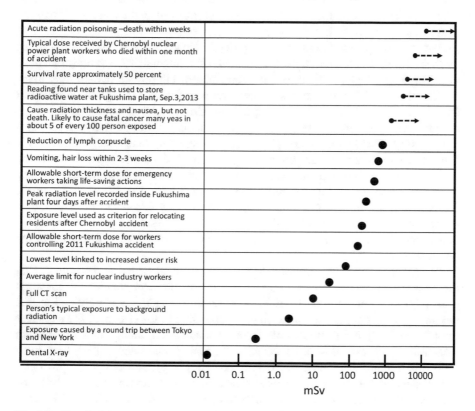

Fig. 2.2 Absorbed dose equivalent given by EQ exposure in natural life and by accident, and the effects of EQ exposure appearing in human beings after the exposure

2.5.4 Conversion of Units Related to EQ Exposure; Intensity (Bq), Absorbed Dose (Gy), Absorbed Dose Equivalent (Sv), and Effective Dose (Sv)

Once the intensity of EQ at the location of a substance is determined in Bq unit, the absorbed dose in the substance can be determined, if the types and energy of EQ and character of the substance are known. In case the substance is a human body, the absorbed dose can be converted to the absorbed dose equivalent (Sv). It should be noted that the absorbed dose is not necessarily uniform when the volume of the body is large compared to an EQ source. Furthermore, different from γ–photons, α– and β–particles can penetrate only very thin depth. Hence the absorbed energy given by their exposure could be very much localized and might be hard to represent their absorbed dose in kg base. Therefore, the comparison of the effects of exposures to α– and β–particles and to γ–photons are not easy with Sv, even though the conversion from Gy to Sv can be made with using the radiation weighting factor (W_R).

Fortunately, or unfortunately, the energy deposition in the external exposure of a human body given by the α– and β–particles (except very high-energy electrons) is limited mostly to skin and the effect hardly appears in tissues or organs in the body.

Different from the α– and β–particles, γ–photons penetrate deep and their energy deposition in a human body is rather uniform if the sources are not very small. Nevertheless, the effect of exposure appears differently depending on the kind of tissues or organs. Hence additional consideration (the tissue weighting factor, W_T) is introduced to convert the absorbed dose to the effective dose as discussed in the previous section.

In any way, conversion from Bq to Gy is possible, considering the type of EQ, their energy, and characters of the substance. For the determination of absorbed dose (Gy) in a human body by EQ exposure, often made is a simplification that the human body consists of water. Since examination of the effects of EQ exposure on a human body is hardly possible from a humanitarian standpoint, EQ exposure tests using a phantom that simulates a human body (tissues and organs) have been done to determine absorbed dose to the human body (Gy). Changing the type of EQ, radiation weighting factor can be also determined to convert from Gy to Sv. Still, the tissue weighting factor (W_T) is necessary to consider the different characteristics of the tissues or organs.

Since the EQ exposure of a human body is mostly caused by EQ sources in air, a conventional conversion from Bq to effective dose (Sv d^{-1}) is proposed when the EQ sources in air are ingested as,

$$A = C \times S \times K_a \times Q \times T \tag{2.17}$$

A Effective dose (mSv).
C Intensity of EQ in air (Bq cm^{-3}).
S Coefficient of outside staying time (=$(S_1 + f_c \times S_2)/24$ h).
S_1 Outside staying time; 8 h.

S_2 Inside staying time; 16 h.

f_c Reduction factor; nearly 1/4.

K_a Effective dose conversion factor (mSv Bq^{-1}).

Q Ingestion (cm^3 day^{-1}).

T Ingestion period; 1 day.

This conversion is given on the website of the Institute of Applied Energy (IEA), Japan [5].

Since the effective dose conversion factor K_a is specified by the type of an EQ source, the effective dose can be estimated according to Eq. (2.17). In case of the exposure to γ–photons, which is the major cause of the air exposure, K_a is nearly constant for different energies so that a conventional dosimeter, like a pocket dosimeter, employs a fixed K_a to give the effective dose directly from the intensity (Bq).

In US, Environmental Protection Agency (EPA) provides a tool to estimate one's yearly effective dose from the most significant sources of ionizing radiation in its website [6].

As mentioned several times, it should be noted that for determination of the effective dose, considerable errors and uncertainties are always included. For example, it does not make much sense to distinguish between 6 and 8 μSv. Also, the absorbed dose is different depending on the type of tissues exposed, as shown in Table 2.2. Therefore, for the exposure of lower dose rate, like μSv/h order, the value of the effective dose rate is just a guide. There is no need to worry about such low dose exposure in daily life. If it rains, EQ sources in air fall down with water. Accordingly, the air dose rate certainly rises but is still too low to worry about.

2.6 Shielding and Decontamination

EQ exposure to a human body is caused by EQ sources mostly in air. Unlike contamination caused by agrichemicals or pesticides, there is no chemical methods to convert the EQ source to be non-hazardous. Although disinfection with alcohol or hot water can be used to remove surface contaminants (EQ sources), that does not work for the removal of EQ sources ingested into the body. In addition, once a substance is exposed to EQ and absorbs its energy, it is not possible to return to the state before the exposure. There is no other way than preventing the exposure (reducing the exposed EQ intensity or the absorbed (deposited) energy). If the EQ source is on the body surface, of course, it can be removed by scraping off.

There is one misunderstanding that not a small number of people seem to believe, that is, once a substance (including a human body) is exposed to EQ, it becomes radioactive or radioactivity is transferred from the EQ source to the substance. This is completely wrong. The reason for the misunderstanding seems to consider that the EQ exposure is something like exposure to virus to catch a cold, and the radioactivity could be transported.

Unless EQ sources are not transferred, the radioactivity cannot be transferred. As detailed in Chap. 3, the dominant EQ source is radioisotopes (RIs) which have extra energy in their nuclei. Since the EQ energy emitted from RI is not high enough to destabilize other nuclei, it is very special for an exposed substance to EQ to become radioactive. The substance exposed to EQ absorbs energy as absorbed dose, and the absorbed energy turns finally to heat. Hence, the EQ sources are not transferred from the substance to substance or a person to person.

As described in the next section, EQ exposure to a human body induces some effect (irradiation effects), so any living beings even bacteria and vaccines are influenced by the EQ exposure. If the absorbed dose is large, it can lead to their death. This is the reason that EQ is used for sterilization. Again, the sterilization by the EQ exposure never makes the substance radioactive.

To avoid or reduce the EQ exposure, it is important to know what and where the EQ sources are. Once you know where the source is, the first thing to do is take the distance from it. Second is to try to reduce the absorbed dose. If there is some material between the source and the substance, EQ deposit some or all of its energy into the material before injecting to the substance. If the material is thick enough or heavy enough to fully absorb the EQ energy, no more EQ injects into the substance, which is referred to as shielding. In case of visible light, a black paper or dark screens can shield it, while as seen in Fig. 1.8 in Chap. 1, EQ requires a thick or dense material to be shielded. The heavier the shielding material, the more energy is absorbed in the material, so that EQ becomes difficult to penetrate.

Thus, to reduce EQ exposure, heavy materials like lead (Pb) whose density is 11.34 g cm^{-3} inserted between an EQ source and a substance or a human body. Still, its thickness of more than 10 cm is required to fully shield γ–photons emitted from ^{60}Co, one of the most famous RIs. Lead grass, yellowish transparent, is often used as a shielding window in a grove box to handle radioactive materials.

Again, the shielding means both reducing the intensity and energy of EQ. Therefore, those who work in intense EQ field, they should take a heavy protection wear, like a spacesuit, not only for shielding but also for avoiding the contamination of possible RE sources in atmosphere.

If an EQ source adhered to a person or materials around him, he should remove the source. This is called **decontamination**. Temporarily blowing dry air to blow off the source is useful but not desirable because the source is only transferred to other places. Suction collection with something like a vacuum cleaner, wiping off with a damp cloth (called smearing), washing off, etc. are effective for the decontamination.

Those EQ sources solvable in water can be removed by washing with water. However, if an EQ source is very intense, dilution of the water used for the decontamination becomes necessary. Radioactive isotopes dissolved as ions in water, like cesium (^{137}Cs) and iodine (^{131}I), can be removed by an ion exchange method or isotope replacement with using their stable isotopes. However, if ions with different charge states are mixed, like Cs^{+} and IO^{3-}, simultaneous removal of them by the ion exchange becomes difficult. In the case that the EQ sources are small particles suspended in water, filtration can be effective, although the effect depends strongly on the method of the filtration. A simple filtering like a simple water purifier may work but is not very effective.

2.7   Effects of EQ Exposure on a Human Body

Figure 2.2 shows absorbed dose equivalent given in natural life and the effects of EQ exposure appearing in human beings after the exposure. The absorbed dose equivalent that clearly induces some observable effects in a human body is 500–1000 mSv or higher. When the human body was exposed to EQ over such high level, symptoms (referred to as acute radiation sickness) appear such as burns and the decrease of white blood cells. This is for one short time exposure. The appearance of the exposure effect is very likely differ depending on the way of the EQ exposure, even the absorbed dose equivalent is the same, for example, longer time with lower dose, periodical exposure, one shorter time with higher dose. Probably the effect is the highest for the last case, although it is not well known how different the exposure effects among the different ways of the exposure.

Allowable absorbed dose equivalent for professional workers is 100 mSv for one year with which exposure no appreciable symptom will appear. For ordinary people, the allowable absorbed dose equivalent is set to be 1 mSv y^{-1}, which is 100 times less than that for the professionals and even less than the annual absorbed dose equivalent in nature of about 2.4 mSv.

The most concerning effect given by the EQ exposure is canceration. In Japan, currently, about 350,000 people are dying of cancer in a year. With the increase of life expectancy, the canceration rate and the number of cancer death are elevating. However, there are only a limited number of cancers for which the cause can be identified. It is well known that smoking increases the canceration rate. Compared to nonsmokers, the rate for smokers is higher about 4.5 times for lung cancer and 30 times for laryngeal cancer.

EQ exposure does not come up canceration immediately. As shown in Fig. 2.2, absorbed dose equivalent of over 500–1000 mSv increase the canceration rate, though quantitative estimation is not possible for lower absorbed dose equivalent. Moreover, it is not yet known whether there is a threshold leading to the canceration. If a person exposed to 1 mSv in a year died from the canceration, it would be difficult to distinguish the cause of the death from other possible causes, such as carcinogenic agents in foods, environmental conditions, diet, and individual differences. The influence of pesticide residues in the environment may be stronger than the EQ exposure. Even the stress of thinking of the canceration could induce cancer.

In contrast, there is a claim that the exposure to very small absorbed dose equivalent is "good for health", which is said to be radiation **hormesis**, i.e., "the exposure of small absorbed dose equivalent activates metabolism and enhances health." Actually, hot spa in which radium (Ra) or radon (Rn) is artificially added is used as balneotherapy. Believing the positive effects of EQ exposure on health could induce a psychological effect to improve health conditions. For example, many people believe that stress-relieving is effective to keep their health.

At the accident of the Fukushima nuclear power plant, various EQ sources were released and dispersed in the atmosphere nearby. Nevertheless, in those areas where the air dose rate is within 100 times of the natural level, the exposure effects of the air dose would hardly appear and the probability of canceration caused by the exposure does not exceed that caused by other factors in the surroundings. Instead, the stress caused by the exposure (even if the absorbed dose was low) and evacuation from the contaminated area are more likely to infect diseases including canceration. As shown in Fig. 2.2, a round trip by air flight between Tokyo and New York gives absorbed dose equivalent of approximately 0.2 mSv. Although a businessman who makes 10 times of the round trip flight is exposed to 2 mSv, no one claims canceration caused by the flights. Probably other factors like jet lag and business-induced stress are more influential on his health. It should be mentioned that the recovery power (encouraged by vitality or sprit) is stronger in those who are actively working in daily life than in those depressed.

It is said that a pleasant sleep promotes the discharge of amyloid β, a substance that contributes to Alzheimer's disease, from the brain, and delays the onset of Alzheimer's disease. If one worried too much about the EQ exposure to depress the daily life, he would not live a happy life and would suppress his recovery power. Instead of being overly concerned about the EQ exposure, one should concern his daily life to avoid a physical or mental disorder due to different causes. There are a few people who seem to have too many concerns about the hereditary effects of the EQ exposure on their children, but their "fearful appearances" could cause the feeling of scary or stress on their children, giving bad influence.

It should be noted that the effects of EQ exposure with low absorbed dose equivalent on a human body can be evaluated only probabilistically. The evaluation is not for the individual but for large numbers of people. It is expressed as "the probability to appear a disease caused by the EQ exposure". For example, the probability of 0.1% means one person out of 1000 people exposed to the same absorbed dose develops a disease. To confirm this probability more than 10,000 people at least should be subjected to the exposure test. Even so, it is not possible to identify the individual developing the disease. The effects of EQ exposure to a substance appear quite different depending on whether the substance is inorganic- or organic-materials, or living beings. For the living beings, the effects greatly depend on their size or mass. As detailed in Chap. 4, the higher living beings are more complex in their structure and the larger their size, more easily they are affected by EQ exposure. Table 4.1 in Chap. 4 compares the lethal doses for various living beings. It is easy to see how human beings are more susceptible than microorganisms.

Again, the difficulty is that although the probability of appearance of some disease caused by EQ exposure can be estimated, for example, as 0.1%, which means that one out of 1000 people is affected, it cannot be determined who is the affected one among 1000 but anyone can be the one. In traffic accidents, the probability of encountering an accident is rather high. However, the society seems accepting it as unavoidable and compensates for the damage caused by the accidents with insurance money that is paid by people getting benefits to use cars. On the contrary, any risk caused by EQ exposure does not seem accepted even if it is very low. Some people say that the effects of the EQ exposure should be reduced to zero. However, that is not possible because risk owing to natural radiation never be zero. Any thermal power plant for electricity generation burns fossil fuel and emits carbon dioxide and other hazardous substances including RIs like tritium. Utilization of any power source always accompanies some risk. Facing to shortage of fossil fuels, nuclear energy is expected to be an alternative of them. To accept the nuclear energy as a power source, the nuclear energy system should be as safe as possible. At the same time, the society should accept the concept that those who are getting benefits should take the accompanying risk, as an insurance system for traffic accidents. That is not possible without correct understanding of "radiation", which is the main purpose of this book. This point is discussed again in Chap. 9.

References

1. http://www-sk.icrr.u-tokyo.ac.jp/sk/index-e.html
2. http://www.pref.saga.lg.jp/kiji00355964/3_55964_53422_up_0a86c1li.pdf [Reprinted with permission]
3. ICRP Publication 103, *The 2007 Recommendations of the International Commission on Radiological Protection*, ICRP Publication 103. Ann. ICRP 37 (2.4)
4. European Commission, Director General for Health and Consumers, Health effects of security scanner for passenger screening (based on X-ray technology), 3.5 Dosimetric aspects, pp. 22–27, EU Publication, Dec. 2012, ISSN 1831-4783, ISBN 978-92-79-26316-3. https://doi.org/10.2772/87426
5. http://www.iae.or.jp/great_east_japan_earthquake/info/appendix2.html
6. https://www.epa.gov/radiation/calculate-your-radiation-dose

Chapter 3
Sources of Energetic Quanta (EQ) (Radiation Sources)

Abstract Sources of EQ (Energetic Quanta) or radiation sources are introduced with separation of radioisotopes (RIs), radiation from the sun, artificial EQ sources including accelerators, X-ray generators, laser, and nuclear reactors. Accidental EQ emission, in particular, at the nuclear accident in Fukushima power plants caused by the earthquake is also described. Since radioisotopes are mostly the EQ source in nature, the EQ emission from RI is detailed.

Keywords Artificial sources · EQ sources · Exposure · Nuclear reactor · Radioisotope (RI) · Accelerator · Laser · X-ray generator

3.1 Radioisotope (RI)

3.1.1 Stable and Radioactive Isotopes

Figure 3.1 shows the electronic structure of copper atom (Cu). In an atom, its nucleus consists of protons with the same number as its atomic number, and neutrons about two times of the atomic number is surrounded by electrons with the same number as the atomic number. Although protons are repulsive with each other due to the Coulomb force, at very near distances they attract each other due to nuclear force. Inclusion of neutrons further stabilized the nucleus also by the nuclear force. The nuclear force works among neutrons and protons with the assistance of meson, and the presence of protons and nearly twice the number of neutrons make the nucleus the most stable. If the number of either protons or neutrons is too large, the nucleus becomes unstable and emits γ-photons, electrons, positive electrons, protons, He, etc. to become a stable isotope different in atomic or mass numbers from original one, An atoms having the same atomic number (or number of protons in the nucleus) but a different number of neutrons (and thus different mass numbers) is called an isotope and represented as $_A^M Z$ with Z as the notation for the atom and A and M are its atomic and mass numbers, respectively. In Cu, there are two stable isotopes in nature with its atomic number of 29 but different mass numbers of 63 and 65 (represented as $_{29}^{63}$Cu and $_{29}^{65}$Cu) with abundance of 69.17% and 30.83%, respectively.

© Kyushu University Press 2022
T. Tanabe, *Radiation: An Energy Carrier*,
https://doi.org/10.1007/978-981-19-1957-2_3

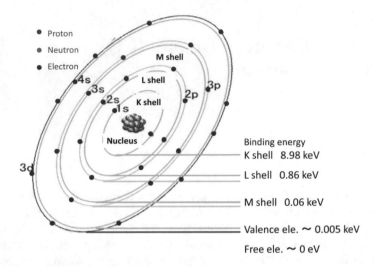

Fig. 3.1 Electronic structure of Copper (Cu) atom

In most of atoms naturally present on the earth, their atomic numbers are between 1 and 92 with several isotopes for each. Most of isotopes are stable (stable isotopes) to continue to exist as they are. In nature, there are unstable isotopes having extra energy in their nucleus to disintegrate to stable isotopes emitting the extra energy as energetic quanta (EQ) either particles or electromagnetic waves, which are referred as radioisotopes (RIs). RIs such as tritium (^{3}H or T), carbon-14 (^{14}C), and potassium-40 (^{40}K) are well known. Although there are two radioisotopes of $^{64}_{29}$Cu and $^{67}_{29}$Cu, their lifetime is so short (12.7 h and 61.9 h, respectively) that they do not exist in nature.

Since the physical and chemical properties of isotopes with the same atomic number are very similar, it is not easy to distinguish stable isotopes with the same atomic number, so as RIs. Slight differences in physical and chemical properties among the isotopes with the same atomic number are called isotope effects, which are mainly due to differences in masses. Since RIs are easily detected by measuring EQ emitted at their disintegration, they are often used as a tracer. In the tracer technology, the behavior of a particular element in a substance is tracked adding a very tiny amount of its RI. In biology and medical field, tritium (^{3}H), carbon (^{14}C), and phosphorus (^{32}P) are often used as a tracer to examine the behavior of water (H_2O), carbon oxide (CO and CO_2), and phosphate (PO_4^{-3}), respectively.

In nature, except cosmic rays, EQ is emitted from RIs, or any EQ sources include RI. In recent days, EQ and some RI can be artificially generated.

3.1.2 Emission of EQ from RI (Disintegration of Nucleus)

Figure 3.2 schematically shows three typical patterns of disintegration of RIs with emission of EQ. As discussed in Chap. 2, they are referred to as (1) α-decay, emitting an α-particle (He^+ or He^{2+}), (2) β-decay, emitting a β-particle, and (3) γ-decay emitting a γ-photon. The energy release from RIs is halved over a specified time called as a half-life (see Sect. 2.2.2 in Chap. 2).

Normal hydrogen (called light hydrogen) is an atom consisting of one proton as nucleus with atomic number 1 and one electron and denoted as ^1_1H or simply H. Hydrogen has another two isotopes, deuterium of one proton and one neutron, ^2_1H or simply D, and tritium of one proton and two neutrons, ^3_1H or simply T. Among the three isotopes, T has excess energy of 0.0186 MeV in its nucleus and disintegrates to 3-Helium (^3_1He) emitting a β-particle as shown in Fig. 3.3. Since T emits an electron as the β-particle, the atomic number of T increases by one to be He after its disintegration. Since the excess energy is distributed between the electron and helium, the maximum energy of the electron is 0.0186 MeV (18.6 keV). Figure 3.3 is so-called the nuclear disintegrate diagram of T, which shows T disintegrates to ^3_2He emit an electron with the half-life of 12.33 years. (Since a normal electron has negative charge, it is denoted as β⁻ and distinguish from positively charged one, positron, β⁺.

Fig. 3.2 Three typical patterns of nuclear disintegration with EQ emission. I, A, and M represent name of an element, its atomic number, and mass number, respectively

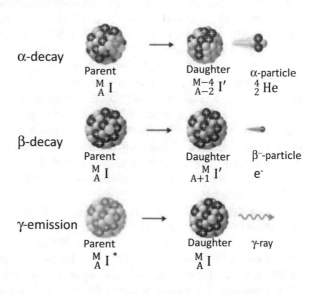

α-decay

Parent
$^M_A I$

Daughter
$^{M-4}_{A-2} I'$

α-particle
$^4_2 He$

β-decay

Parent
$^M_A I$

Daughter
$^M_{A+1} I'$

β⁻-particle
e⁻

γ-emission

Parent
$^M_A I^*$

Daughter
$^M_A I$

γ-ray

Fig. 3.3 Decay scheme of tritium

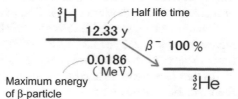

^3_1H Half life time

12.33 y

β⁻ 100 %

0.0186
(MeV)

Maximum energy
of β-particle

^3_2He

3.1.3   RI in Nature

Any stable isotopes can be transferred to RI, if extra energy is given in its nuclei and the created RI disintegrates to a stable isotope emitting the extra energy. Although most of atoms on the earth today consist of stable isotopes, there are a few RIs. At the time of the earth's birth, the entire earth had enormous amounts of energy, in other words, there were various RIs that emitted energy. After 4.6 billion years from the birth, most of RI initially existed have decayed out but still, some RIs remain. Figure 1.5 in Chap. 1 shows dominant RIs on the present earth, such as ^{40}K, ^{87}Rb, ^{147}Sm, ^{176}Lu, ^{187}Re, ^{232}Th, ^{238}U, and ^{235}U. The last three are used as fuels of nuclear reactors. Since these RIs have very long half-lives, they still remain. ^{40}K is well known to be included in various foods. Figure 1.7 in Chap. 1 shows how ^{40}K is distributed in the food (root vegetables) as mappings of emitted EQ from ^{40}K.

In addition to these long-life RIs, there are some RIs produced by nuclear reactions of stable isotopes with cosmic rays like T and ^{14}C mentioned above. In addition, T is/was artificially generated in nuclear reactors and nuclear bomb tests conducted in the 1950s and 1960s. Before the second world war II, T concentration in atmosphere had been nearly constant, keeping the balance between the generation by the cosmic ray and the loss by its decay. However, huge amount of T was produced and released in atmosphere by the nuclear bomb tests to increase its concentration by about 100 times. The concentration peaked in 1963 when PTBT (Partial Test Ban Treaty) became effective. Afterward it has decreased following exponential decay with its half-life (13.6 year). The concentration of ^{14}C had also increased with the nuclear tests and decayed after PTBT but a little slower than T. See Fig. 1.6 in Chap. 1.

^{14}C is used in a method of determining the age of specimens from extinct animals and plants, called carbon dating. While living beings are alive, the concentration of ^{14}C in them is kept constant because carbon in the air is taken in and out by metabolism. However, after their death the concentration of ^{14}C decreases following its half-life time due to the stop of their metabolism. This makes it possible to determine how many years have passed after their death or stopping of life activity. Archaeology uses this to age excavated wood, buildings, furniture, furnishings, etc. as the carbon dating or radiometric dating. Its details are given in Sect. 7.5 in Chap. 7.

Hot springs containing Radium and Radon are used as known as balneotherapy. Hot spring containing 226-Ra (^{226}Ra) with its concentration of 10^{-8} g L^{-1} is referred to as radium hot spring and that containing 222-Radon (^{222}Rn) as the decay product of ^{226}Ra with its activity 74 Bq L^{-1} as radon hot spring. Stimulation given by EQ exposure with a small absorbed dose equivalent is believed to be good for health. This is referred to as radiation hormesis. Ra and Rn are some of the decay products of uranium (U) and thorium (Th) as described below.

Some natural ores such as monazite and phosphorite contain ^{235}U, which is used as nuclear reactor fuel. They also contain ^{232}Th and ^{238}U, which can be reactor fuels in future. The reason for existence of high air dose rate areas given in Table 1.2 in Chap. 1 is that the areas contain these ores in their ground. Coal also contains Th and U with their radiation intensity of about 0.002 to 0.02 Bq g^{-1}. Therefore, the exhaust gas from thermal power plants contains radon (Rn), and the ash contains U and Th and even radium (Ra). The absorbed dose equivalent at the ash storage site of a thermal power plant is estimated to be up to 0.2 mSv y^{-1}. This is about one-tenth of the natural exposure dose.

3.1.4 EQ Exposure of Human Body in Nature

As described above, as EQ sources in nature there are natural RIs, cosmic rays, and RIs produced by the cosmic rays, and any life beings are exposed to EQ from these sources. The absorbed dose equivalent caused by the cosmic rays for human beings is 0.4 to 3 mSv per year. Because the shielding effect of air decreases with increasing altitude, the absorbed dose equivalent doubles with every 1500 m going up in the sky. Accordingly, the absorbed dose equivalent given by an intercontinental flight between Tokyo and New York over 10,000 m high becomes 0.2 mSv.

In nature, there are several EQ sources other than the air exposure. They are foods containing ^{41}K, soils and ores including ^{232}Th and ^{238}U, phosphorus fertilizer made of mineral phosphate, ashes and exhausted gas including T and other RIs from thermal power plants, residues of nuclear bomb tests, and others. In total, natural annual absorbed dose equivalent is around 2.4 m Sv^{-1} as shown in Table 3.1. However, as shown in Table 1.2 in Chap. 1, the absorbed dose equivalent in the world is widely scattered depending on locations with the difference of more than 100 times between the maximum and the minimum. In Japan, the highest absorbed dose equivalent (20 times) is observed at Ningyo Pass in Okayama prefectural where U ores were minded.

By the way, all stars radiate energy as EQ produced by fission or fusion reactions inside of them. The sun is the energy source for the earth. As described in Chap. 1, the atmosphere of the earth shields dangerous EQ emitted from the sun. Photons with higher energy than ultraviolet light, which are very dangerous, hardly come to the ground surface so that human beings and other living beings can continue to exist on the earth.

After around 10 billion years since the birth of the earth, lives were born. There is a theory that EQ contributed to the birth of the life. Because EQ can cause chemical reactions which are not possible with conventional chemical reactions. For example, EQ irradiation to inorganic materials consisting of carbon, hydrogen, nitrogen, phosphorus, and so on causes some chemical reactions among them generating some complex organic molecules or life forms. EQ also likely contributes evolution of living beings through mutations caused by the EQ exposure. Since radiation level in the early days of the earth was significantly higher than it is now, it should give not a small influence, whether positive or negative, although it has not been clarified how and what was the influence. Because the living beings have been continuously exposed to a certain level of absorbed dose equivalent, they seem to get some ability of radiation resistance. In other words, even if cells or DNA were damaged by EQ exposure, they could repair by themselves, or use a self-recovery system, which is discussed in detail in Chap. 8

Table 3.1 Averaged annual absorbed dose equivalent given by various EQ in nature (World average)

EQ Source	Averaged annual absorbed dose equivalent (mSv)
Cosmic rays	
Ionic radiation and/or photons	0.28
Neutral	0.1
RIs generated by the cosmic rays	0.01
Subtotal	0.39
Air exposure	
Outdoors	0.07
In doors	0.41
Subtotal	0.48
Internal exposure due to inhalation	
Uranium and Thorium Series	0.006
^{222}Rn	1.15
^{220}Rn (Thron)	0.1
Subtotal	1.26
Foods	
Internal exposure due to ingestion of food inhalation	
^{40}K	0.17
Uranium and Thorium Series	0.12
Subtotal	0.29
Total	2.4

3.1.5 EQ Emitted from 131-Iodine and 137-Cesium and Their Exposure Effects

Figure 3.4 shows disintegration schemes of 131-Iodine (^{131}Cs) and 137-Cesium (^{137}I) which are major RIs as EQ sources released from the Fukushima nuclear power plant. ^{131}I first disintegrates with β-decay emitting an electron becoming metastable 131-Xenon (^{131m}Xe) with a half-life of 8.02 days (see Fig. 3.4a). ^{131m}Xe is unstable with two different energy states and emits its extra energy as γ-photons almost instantaneously. The probability to take which decay processes is indicated by %. The main energy of EQ emitted by the disintegration are those with energy of 0.807 meV (0.39%), 0.334 meV (7.23%), 0.606 meV (89.6%) as β-particles, and 0.0802 meV (2.62%), 0.284 meV (6.12%), 0.364 meV (81.5%), and 0.637 meV (7.16%) as γ-photons. Numbers in parentheses (%) indicate the probability of each emission. The total emission intensity of EQ is 4.6×10^{15} Bq kg^{-1} of ^{131} I.

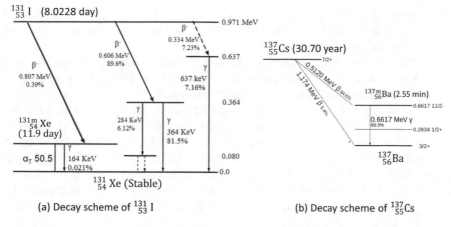

(a) Decay scheme of $^{131}_{53}$ I   (b) Decay scheme of $^{137}_{55}$ Cs

Fig. 3.4 Decay schemes of **a** $^{131}_{55}$ I and **b** $^{137}_{55}$ Cs

In case of the external exposure, most of the energy of β-particles is absorbed in skin resulting in symptoms of burns and skin cancer. However, the risk of canceration of body tissues is lower than given by the exposure to γ-photons. For the internal exposure, thyroid cancer caused by ^{131}I is concerning, because I is easily accumulated in the thyroid, and β-particles emitted from ^{131}I accumulated in the thyroid deposit their energy directly in it. If 2.2×10^{-9} g of ^{131}I equivalent to 10^4 Bq is ingested, resulting internal absorbed dose equivalent is 0.22 mSv y^{-1}. The main pathway through which a person intakes iodine is grass → cows → milk → people as food chain. This transportation proceeds quickly, and the concentration of radioactive iodine in milk peaks three days after it is deposited on the pasture. The effective half-life for its removal from the pasture is about 5 days. If 1000 Bq of ^{131}I is deposited on the pasture of 1 m^2, about 900 Bq is estimated to be transferred into 1 L of milk. Suppose 10% of ^{131}I contained in the milk is accumulated in the thyroid, the ingestion of 1000 L of milk results in 10^4 Bq as described above. For comparison to the internal exposure, the external exposure with γ-photons is considered. If one is exposed to γ-photons emitted from their source of 10^8 Bq (equivalent to ^{131}I of 0.22 mg) at 1 m apart, the absorbed dose equivalent is 0.0014 mSv d^{-1}.

^{137}Cs disintegrates to ^{137}Ba emitting either β$^-$ electrons with a half-life of 30.7 years, while 94.4% of the disintegration is through the metastable atom of ^{137m}Ba emitting 0.512 meV β$^-$ electron which suceedingly disintegrates to ^{137}Ba emitting 0.662 meV γ-photons with a half-life of 2.6 min and 5.6% emitting 1.174 meV β$^-$ electron (see Fig. 3.4b). The emission intensity of EQ is 3.2×10^{12} Bq kg^{-1} per kg of ^{137}Cs. This value is much smaller than that of ^{131}I. This is because ^{137}Cs has a much longer half-life than ^{131}I, i.e., lower energy release in unit time. If 10^4 Bq of ^{137}Cs was ingested, the effectively absorbed dose equivalent should be 0.13 mSv. In contrast, the external exposure to γ-rays from 10^8 Bq (320 mg) of ^{137}Cs at the distance of 1 m, the effective dose equivalent should be 0.0019 mSv per day. Cs also have another RI of ^{134}Cs, which decays with a short half-life of about two years. Therefore,

^{137}Cs were detected immediately after the Fukushima nuclear accident (See Fig. 6.2 in Chap. 6), while after 10 years of the accident its intensity has become very low, while ^{134}Cs is appreciable.

Just after the accident, exposure to EQ from ^{131}I and ^{137}Cs was concerned, because both have a short half-life compared to various long-lived ones and the emission intensity and therefore the energy emission rate is high.

The half-lives of other radioactive fission products emitted at the accident are summarized in Table 3.2. EQ intensity around the Fukushima nuclear power plant significantly increased by three times of hydrogen explosions. Then the intensity decayed rapidly owing to firster decays of short live RIs such as ^{131}I with a half-life of 8.05 days. Afterward, it is gradually decreasing owing to longer life isotopes like ^{137}Cs of 30.7 years. Without further release, radiation intensity will gradually decrease according to the decay of ^{131}I and ^{137}Cs, and then other isotopes with much longer life will become apparent.

Table 3.2 Fission products generated by nuclear fission of ^{235}U

Production rate given by one nuclear fission (%)	Fission products	Half-life
A few %	^{132}Te	3.2 d
6.7896	^{133}Cs → ^{134}Cs	2.065 y
6.3333	^{135}I → ^{135}Xe	6.57 h
6.2956	^{93}Zr	1.53 My
6.0899	^{137}Cs	30.17 y
6.0507	^{99}Tc	0.211 My
5.7518	^{90}Sr	28.9 y
2.8336	^{131}I	8.02 d
2.2713	^{147}Pm	2.62 y
1.0888	^{149}Sm	Non-radioactive
0.6576	^{129}I	1.57 My
0.4203	^{151}Sm	90 y
0.3912	^{106}Ru	373.6 d
0.2717	^{85}Kr	10.78 y
0.1629	^{107}Pd	6.50 My
0.0508	^{79}Se	0.327 My
0.0330	^{155}Eu → ^{155}Gd	4.76 y
0.0297	^{125}Sb	2.76 y
0.0236	^{126}Sn	0.23 My
0.0065	^{157}Gd	Non-radioactive
0.0003	^{113m}Cd	14.1 y

d: days, h: hours, y: years, My: mega years (10^6 years)

3.2   Radiation from the Sun

The sun is the energy source of the earth and supports all lives. In the sun, energy is generated by fusion reactions as shown in Fig. 3.5. Although the reaction passes to generate the energy are complicated, as a whole, four protons and two electrons are fused to be ^{4}He and the generated energy of 26.65 MeV is carried by ^{4}He, γ-photons, and anti-neutrinos as EQ. The fusion reactions proceed very slowly inside the sun. Hence EQ generated by the reactions is not emitted directly from the surface of the sun, but is converted into lower energy EQ, i.e., long-wavelength X-rays, ultraviolet light, and visible light.

Table 3.3 shows the types of EQ and their distribution rates emitted from the solar surface. Most of the energy is emitted as ultraviolet, visible, and infrared lights. Rainbows tell us that sunlight is made up of light of various wavelengths. Figure 3.6 [1] shows wavelength distribution of photons (electromagnetic waves), which is referred as the solar spectrum. The spectrum is dominated by a visible light in the wavelength ranging from 400 nm to 10,000 nm with the highest intensity at around 500 nm. The spectrum is similar to the radiation spectrum from a blackbody at 5800 K which is defined as the surface temperature of the sun. Since the earth is covered by the atmosphere, some of the sunlight is absorbed by various molecules in the atmosphere to change the wavelength distribution from the sunlight outside of the atmosphere to that at the ground surface as indicated in Fig. 3.6.

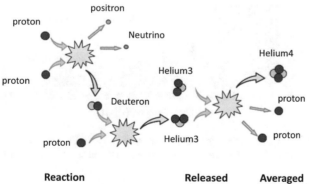

Reaction		Released energy	Averaged reaction time
P + P	→ D + e$^+$ + v	+ 0.4 MeV	1.4 x 10^{10} years
e$^+$ + e$^-$	→ 2γ	+ 1.0 MeV	10^{-19} sec
P + D	→ ^{3}He + γ	+ 5.5 MeV	5.7 sec
^{3}He + ^{3}He	→ ^{4}He + 2p	+ 12.85 MeV	10^8 years
	Total		
4p + 2e$^-$	→ ^{4}He + 6γ + 2v	+ 26.65 MeV	

Fig. 3.5 Fusion reactions occurring in the sun

Table 3.3 Energy and wavelength of electromagnetic waves and particles emitted from the sun

Name	Wavelength	Emitted fraction
γ-photons	~10 nm (~0.1 MeV)	Very small
X-rays	10–400 nm (100–1 keV)	Very small
Ultraviolet	~0.4 μm (~6 eV)	~7%
Visible	0.4–0.7 μm	~47%
Infrared	0.7–100 μm	~46%
Micro-wave	100 μm~	Very small
Nyutrino		Negregiblly small
α-particles, β-particles, Other elementary particles		Emitted from sloar flare, but hardly coming to the ground surface

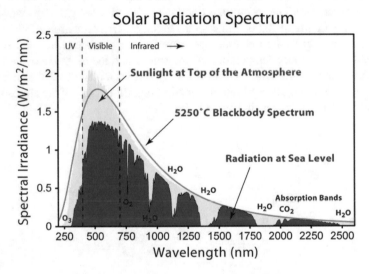

Fig. 3.6 Solar radiation spectra at the top of the atmosphere and at the sea level showing some losses due to the absorption by water and carbon dioxide (in [1], reused with permission)

It should be also noted that ultraviolet and shorter-wavelength light less than 300 nm (10 eV), which is dangerous for living beings, hardly reaches the ground surface. This allowed the survival of living beings and their evolution on the earth. It is well known that in some areas in Australia where the intensity of ultraviolet light is higher than in other areas and the incidence rate of skin cancer is extremely high. The cause is less absorption by water in the atmosphere due to dry air in areas such as deserts and savannahs. The exposure to ultraviolet light is exactly the EQ exposure.

3.3 Nuclear Reactor

In nuclear fission reactors, 235-uranium (^{235}U) is used as their fuel. When a neutron (n) is incident on ^{235}U, it absorbs the neutron and becomes an unstable isotope of $^{236\,m}$U. Then $^{236\,m}$U fissions into two nuclei named fission products (FPs), $^{A}Z_1$ and $^{B}Z_2$, emitting neutrons and γ-photons as

$$n +^{235} U \rightarrow^{236m} U \rightarrow^A Z_1 +^B Z_2 + Nn + \gamma's \qquad (3.1)$$

and schematically shown in Fig. 3.7 [2], where A and B are mass numbers of FPs of Z_1 and Z_2. The energy released by this fission reaction (E) is given by the famous Einstein's equation as the difference between the total masses of the left side and the right side (Δm),

$$E = -\Delta m \times c^2 \qquad (3.2)$$

where c is the speed of light. FPs are not specified but they are two, one with mass numbers of less than 100 and the other larger than 100 as shown in Fig. 3.8 [3]. Also, depending on what the generated FPs are, the number of neutrons generated (N) varies as

$$N = (1 + 235)-(A + B). \qquad (3.3)$$

Consider a nuclear fission reaction, as an example,

$$n +^{235} U \rightarrow^{144} Ba +^{89} Kr + 3n + \gamma \text{ - photon,} \qquad (3.4)$$

which generates 173 MeV of energy. The generated energy is distributed to two FPs of ^{144}Ba and ^{89}Kr, three neutrons, and γ-photons. There are various

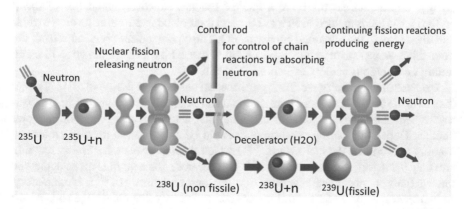

Fig. 3.7 Schematic drawing for fission reactions of ^{235}U (in [2], Reprinted with permission)

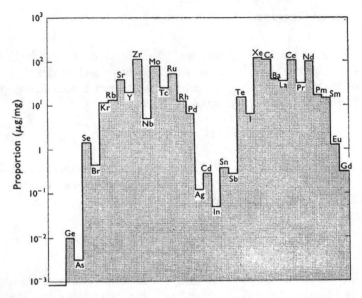

Fig. 3.8 Mass distribution of fission products (FPs) produced by fission of ^{235}U (Reprinted from [3] with permission from Royal Society of chemistry)

types of fission reactions, and the generated energy varies slightly. On average, 180 MeV of energy is released per one fission. The generated energy is distributed to $^{A}Z_1$ and $^{B}Z_2$ (168 MeV) and neutrons (4.8 MeV) as their kinetic energy and γ-photons (7.5 MeV). Most of the generated FPs store excess energy as RIs and some are stable isotopes such as ^{149}Sm and ^{157}Gd. Table 3.2 shows FPs generated by the fission of ^{235}U with their generation probabilities and half-lives. The table shows that the generation probability of ^{133}Cs (^{134}Cs) and ^{135}I, of which disintegration is discussed in the previous section, are quite high.

The generated energy of 180 MeV by the fission reaction is converted to thermal energy to generate electricity in a nuclear reactor. However, about 8% of them, or 14.6 MeV, are remained in FPs as RIs, and their disintegrations according to their half-lives release the remained energy as EQ. The continued energy release from the Fukushima power plants after the accident is caused by these radioactive FPs and requires continuous cooling even after 10 years.

Because the number of neutrons generated in one nuclear fission of ^{235}U is larger than two, the presence of more than a certain amount (critical amount) of ^{235}U causes a so-called chain nuclear reactions in which extra neutrons cause reactions one after another. Depending on the conditions, it could be an explosion used as an atomic bomb. In natural U, the majority is ^{238}U which does not cause the fission reaction and only 0.7% is the fissile isotope of ^{235} U. Therefore, the natural U alone cannot be nuclear fuels nor make a nuclear bomb, i.e., isotopic enrichment of ^{235}U is mandatory. In a nuclear reactor, fission reactions are controlled by using controlled rods which contain neutron absorbers such as ^{10}B or ^{113}Cd as shown in Fig. 3.7, i.e., controlling the number of neutrons available to the chain reactions. It should be noted that ^{238}U absorbs a neutron to become ^{239}Pu, which is a fissile isotope like ^{235}U and hence used as a fuel for a fast breeder reactor.

The neutrons generated in fission reactions have too much energy to continue the chain reactions and hence their energy is reduced in decelerator or moderator in which the energy of neutrons carried as kinetic energy is reduced by slowing down their velocity, which is the reason for referring the decelerator or moderator. (See Fig. 3.9a.) In most cases, normal water (H_2O) is used as the moderator. Hence a power reactor using H_2O as the moderator is referred to as a light water reactor. As shown in Fig. 3.9b [4], uranium dioxide is used for the fuel of nuclear reactors and contained in a sheath made of zirconium alloy named Zircalloy to make a fuel pin or fuel rod. The fuel pins are assembled with around 10×10 to be a fuel assembly for

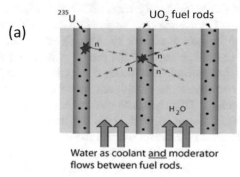

(a)

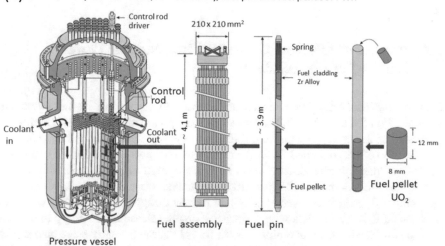

Fig. 3.9 a Deceleration of neutrons by water and b structure of PWR pressure vessel, fuel assembly, fuel pin, and fuel pellet (Reprinted from [4] with permission from Kyushu Electric Power CO., INC., Japan)

example. The fuel assemblies are spread in a reactor and cooling water also works as a moderator flow through gaps between fuel pins. Since the oxidation of the sheath does not proceed at the reactor operating temperature, FPs generated in the sheath are contained in the sheath or the fuel rod.

The cause of the hydrogen explosion at the Fukushima nuclear power plant is oxidation of the sheath (Zircalloy) of the fuel pins,

$$Zr + 2H_2O \rightarrow ZrO_2 + 2\,H_2 \tag{3.5}$$

Although the reaction does not proceed at the operation temperature of the reactor, loss of cooling water caused by the earthquake made the temperature of the fuel pins very high. Consequently, Zr reacted with water vapor in the reactor vessel and massive H_2 as the product of the oxidation was released. In order to avoid the oxidation, any fission reactor equips Emergency Core Cooling Systems (ECCS) which can supply cooling water in emergency. At the Fukushima nuclear power plant, ECCS worked at first, but it stopped because of the loss of electricity. At high temperatures, the oxidation produces heat so that the temperature of the fuel pins is raised further, which in turn enhances the oxidation and enhance the H_2 production. It can be said that the delay in the decision to cool the reactors with seawater injection is one of the causes of the Fukushima accident. This oxidation also causes the fuel pins to lose their function to enclose the FPs even if the fuel did not melt, and the fuels were exposed to water and accelerated FP release. Even in the water pool for spent fuel storage, the fuel continues to generate heat caused by the decay heat of FPs. Therefore, if water was lost in the pool and part of the fuel pin was oxidized, then FP could be released.

3.4 Release of FPs from the Fukushima Nuclear Power Plant After the Accident

Here is considered the FPs released at the Fukushima Nuclear Power Plant. As shown in Fig. 3.8 and Table 3.2, various types of FPs having high radio-activity are generated in the fission reactors and released at the accident. Among them, Cesium and Iodine (often combined as CsI) are two of the most concerned EQ sources, because of their high production rates as FPs and high radio activities. ^{132}Te was one of the most dominant FPs just after the accident but owing to its short half-life (only 3.2 days) it decayed shortly and mostly disappeared now. ^{90}Sr was also dominant FP. Basically, RIs with a shorter half-life release their energy in a short time, so that the intensity of emitted EQ in Bq is larger in the early stage but rapidly decreases. Therefore, radioactive FPs having short half-life and low melting point (high vapor pressure) were released at the accident and detected as highly intensive EQ emissions like ^{132}Te. More than 10 years after the accident, the dominant FPs detected is ^{137}Cs having a half-life time of about 30 years. If ^{132}Te was detected now, a new nuclear reaction should have occurred. Since no new ^{132}Te has been detected after the accident, it indicates that no additional fission reactions have occurred in the plants. In Fig. 6.3 in Chap. 6, the energy spectrum of the released EQ at the accident of Fukushima nuclear power plant are compared for four days and eight-day after the accident together with the background (before the accident). Above mentioned decaying tendency of released FPs is explained again in Sect. 6.2.3 in Chap. 6.

As mentioned in Chap. 1, after World War II, various radioactive FPs were released and dispersed in the atmosphere all over the world by nuclear bomb tests. Therefore, from middle of '50s to '60s, the intensity of EQ (radiation) in the atmosphere, or air dose rate was 1–2 orders of magnitude higher than that given by natural EQ sources. However, since the Partial Test Ban Treaty (PTBT) came into effect at 1963, the air dose rate in the atmosphere has been decreasing. Even at the nuclear accidents in Chernobyl and Three Mile Island (TMI), the increase of the air dose rate was detected. Nevertheless, general decaying tendency was not altered. At that time Cs were detected far from the accident site such as the site of the Genkai Nuclear Power Plant in Kyushu, Japan. The increase of the air dose rate was also detected in the accident of the Fukushima nuclear power plant. Fortunately, the increase of the air dose rate was not so high to influence public health.

Radioisotopes dispersed in the atmosphere have been decaying. However, owing to different half-life of each RI, their attenuation was significantly different with each other. In particular, those having short lifetime like ^{3}H and ^{14}C were decreased to less than 1/50 and 1/10, respectively, i.e., absorbed dose rate given by them returned to the levels before World War II. Although ^{137}Cs and ^{131}I are also decreasing according to their half-life, their attenuation is slow, for example, only 1/3 of ^{137}Cs disappeared. Unfortunately, 60 years ago or in 1963, the air dose rate was hardly measured. However, taking all RIs released and their half-lives, the air dose rate in 1963 is estimated to be 0.1–0.01 μSv h^{-1}, 10 times higher than the present air dose rate of 0.01–0.001 μSv h^{-1}.

About 60 years after PTBT in 1963, the canceration rate in developed countries has increased an order of magnitude. However, the increase is not due to the air dose but caused by the increase in life expectancy and changes in lifestyle. Therefore, it is difficult to evaluate the after-effects of the exposure. Of course, this does not claim that Fukushima is safe, but at least these observations would be useful to avoid intense anxiety and feelings of panic on the nuclear accidents.

As detailed in Chap. 6, it is possible to find what is happening in the environment by measuring the type of EQ and their energy. Although EQ is invisible, easy to detect and measure. Continuous monitoring of the air dose rate using reliable devices (one of the simplest is available as a pocket dosimeter (see Fig. 6.6 in Chap. 6)) will ensure safety and give a sense of security.

3.5   Artificial EQ Source

The development of science and technology has allowed to create artificial EQ sources using various methods and mechanisms. Although some people seem to believe that natural radiation or EQ in nature is different from artificially created EQ, both are exactly the same, if the kinds and energies of EQ in nature and artificially created ones are the same.

Historically, it was very unfortunate that large amounts of RIs were artificially generated by atomic bombs, and were dispersed in the atmosphere. Accordingly, many people were unintentionally exposed to EQ emitted from them without notice. This seems to give bad impression to many people to use nuclear power for electricity generation. As described in this book, with the development of science and technology, we have come to know that there is no difference between EQ in nature and artificially created ones and that the radiation is consisting of EQ-carrying energy, which can be safely handled to be an energy source like nuclear reactors as described in Sect. 3.3. In the following, as artificial EQ sources, accelerators of ions and electrons, X-ray generators, and laser generators are introduced in comparison to natural EQ.

3.5.1   Accelerator

It is not so difficult to ionize atoms or to make charged ions, such as protons (H^+) and electrons (e^-). As seen in Fig. 1.2 in Chap. 1, if the energy of several eV is given to an atom, electrons in the atom can be released from their binding state and consequently, the atom is ionized. Since the ionization energy is very small in alkali metals, they can be ionized simply by heating to about 2000 °C. Usually, ions are produced by the impact of electrons accelerated above around 10 eV to atoms. Once an atom is ionized to be an ion and electron, both can be accelerated with electrostatic field. In Fig. 3.10, the principle of an accelerator is schematically shown. A filament to produce thermal electrons and positive and negative electrodes are placed in a vacuum or low-pressure gas. When the filament is heated in a vacuum, electrons (called

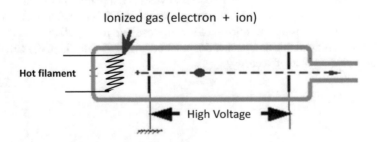

Fig. 3.10   Principle of electrostatic accelerator

thermal electrons) are released. Then, applying an electrostatic potential (voltage) between the two electrodes, the electrons are accelerated to the positive electrode. The accelerated electrons are referred to as electron beams. The X-ray generator described in the next section uses the electron beam. In case of ion acceleration, a small amount of gas is placed in the system or the material evaporates from the filament, the molecules and atoms in the gas are ionized. Since ions have a positive charge, the ions can be accelerated by applying the inversed electrostatic potential for the electron acceleration. Usually, available voltage is about 1 million volts (1MV), electron beams and ion beams accelerated up to 1 MeV can be made. Normal electron microscopes use electron beams accelerated to hundreds of keV. It should be noted that this energy is comparable to the 512 keV electron energy emitted from ^{137}Cs and is even greater than the average electron energy of 190 keV emitted from ^{131}I. Thus, EQ with similar energy to natural EQ or even higher energy can be artificially created in such a simple way (although it is not easy to actually make it) and they are used as electron beams and/or ion beams. Now accelerators that generated even higher-energy beams are available. In high-energy physic laboratories, very high-energy accelerators with energy of GeV, a thousand times larger than 1 MeV. Nihonium created quite recently and named after Japan, or other trans-uranium elements having atomic numbers of larger than 92 can be created with a technique to use the collision of two accelerated heavy ions.

3.5.2 X-ray Generator

X-rays are used in a variety of applications in modern society. Figure 3.11 shows the principle of X-ray generation. First, thermal electrons are generated as described in the previous section (See Fig. 3.10). Then, they are accelerated toward a target plate given a positive potential. When accelerated electrons to higher energy inject into the target, the constituent elements of the target material usually copper (Cu) or tungsten (W) is ionized. As shown in Fig. 3.12 the injected electrons collide with the inner shell electrons to be ejected remaining an electron-hole. Succeedingly, the

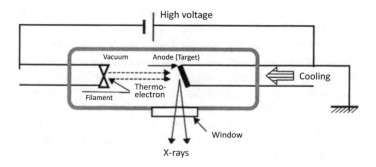

Fig. 3.11 Principle of X-ray generator

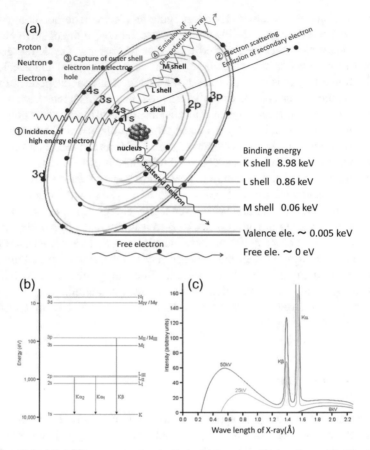

Fig. 3.12 **a** Principle of X-ray generation in Cu by high-energy electron impact, **b** Energy levels of the inner shell electrons of Cu, and **c** Energy distributions of X-rays generated by the incident electrons of 8, 25, and 50 keV. The energy of 8 keV is not high enough to generate characteristic X-rays of Cu- Kα and -Kβ

hole pulls an electron in the outer shell and releases excess energy as an X-ray which is called as the characteristic X-ray. If the energy of the incident electrons is not high enough (8 keV in Fig. 3.12b), it cannot produce the electron-hole in the inner shell so as the characteristic X-ray. As seen in Fig. 3.12b, generated X-rays are not limited to the characteristic X-rays but include lower energy ones, because some of the characteristic X-rays lose their energy during they pass through the target. This is an example showing that the interaction of EQ appears differently depending on their energy. Although X-ray generators use high-energy electrons, X-rays with the same energy instead of the electrons can also generate X-rays. In a device called X-ray fluorescence spectrometer, a sample is irradiated with high-energy X-rays to induce X-ray emission from the sample and measure the energy distribution of the emitted X-rays. If characteristic X-rays of a specified element are detected, one can identify the element included in the sample. X-ray irradiation also causes electron emissions which are used as an X-ray induced photoelectron spectroscopy (XPS).

X-ray projection or X-ray imaging, very important for health care, is using different absorption abilities between heavy and light materials. When taking the X-ray projection of the stomach, barium (Ba) is often taken to enhance the contrast of its X-ray image (photo), because of its heavy property enhancing the X-ray absorption. Since the X-ray projection allows us to know what kind of substance is inside, it is always used for non-destructive analysis. For heavier and/or thicker materials, higher-energy X-rays are required. Since it becomes harder to generate higher-energy X-rays, γ-photons emitted from RIs are used.

In this way, EQ irrespective of its origin, natural or artificial, can be effectively used under controlling and monitoring when its nature (types and energy) is known. Of course, shielding and isolation are necessary to avoid or reduce the exposure of workers as low as possible.

3.5.3 Laser

Recent days, lasers are used in a variety of places and opportunities. Lasers are electromagnetic waves same as γ-photons, X-rays, and light. Different from them, the laser light is easier to control energy, intensity, and time duration (continuous or pulsed), though wavelength region is limited from infrared to violet. Furthermore, the laser light can be focused with a lens, which can increase the power deposited per unit area. As a result, laser cutters, welders, and furnaces have been developed for industrial use, and laser scalpels for medical use for example.

For a continuous laser, the deposited energy is simply given by laser power multiplied with time. It is one of the benefits to use a pulsed laser to get very high power easily. Consider a laser with output energy of 1 J. If it is continuously operated, the output (deposited) power is 1 W, while pulsed with 0.1 s, the power becomes 10 W. Owing to recent technological advancement, the pulsed laser of femtosecond (10^{-15} s) is available. Hence the pulsed laser with output energy of as small as 1 J pulse^{-1} can deposit such high powers of 10^9 W pulse^{-1}, or 1 GW pulse^{-1}. Nowadays, even 10^{15} W (1 PW) laser is available, as a petawatt laser.

Laser researchers have a dream. If the laser power can be increased further, it is possible to directly irradiate nuclei to trigger nuclear reactions, either fusion, fission, or disintegration. Since RIs are unstable with extra energy in their nucleus, they disintegrate emitting the extra energy. Similarly, if the extra energy was introduced into the nucleus of an RI, its nucleus would become more unstable and accordingly disintegrate within a much shorter time than its half-life. As already mentioned in Fig. 1.2 in Chap. 1, there is a certain relationship between the time duration of any physical phenomena and accompanied energy change. That means simple increase of the energy deposition in the nucleus does not necessarily come to the disintegration but to deposit large power deposition within short duration is mandatory. Thus, if an exa-watt laser of which output power is 10^{18} W or 1 J for one pulse of 10^{-18} s is established, that can be used to control nuclear processes, in particular, enhance disintegration or extinction of RIs. This is partly the reason that the title of this book is "Radiation: An Energy Carrier".

References

1. http://commons.wikimedia.org/wiki/File:Solar_Spectrum.png. Drawn by Dr. Robert A. Rohde
2. http://www.athome.tsuruga.fukui.jp
3. B.R.T. Frost, Nuclear fuels. R. Inst. Chem. Rev. **2**, 163–205 (1969)
4. http://www.kyuden.co.jp/company_pamphlet_book_plant_nuclear_genkai_genkai.html

Chapter 4
Irradiation Effects of EQ on Materials (Inorganic- and Organic Materials, and Living Beings)

Abstract The effects of EQ exposure are detailed focusing the evaluation of the effects of low-dose exposure and their mitigation. Based on principal damage formation caused by nuclear collision and electron excitation under EQ exposure, the effects of EQ exposure are summarized separately for inorganic- and organic materials, and living beings. Since organs and tissues of human beings have resilience to EQ exposure causing recovery, the appearance of the effects of EQ exposure is different depending on where or what kind of organs in a human body is exposed.

Keywords Damage · Electron excitation · EQ exposure effect · Nuclear collision · Resilience · Recovery

4.1 Evaluation of the Effects of EQ Exposure

4.1.1 There is no Critical Level in Absorbed Dose to Distinguish Secure and Insecure

Since the Fukushima nuclear reactor accident caused by Tsunami released and dispersed large amounts of radioactive materials, mostly fission products, people living not only in the area but also in some remote areas have been concerned with the EQ exposure and are asking; "Is there any critical level in absorbed dose equivalent (in Sv) to distinguish secure and insecure?" In the case of lower absorbed dose equivalent (approximately less than 100 mSv y^{-1}), it is not possible to identify the threshold that causes health hazards to an individual. Of course, the lower absorbed dose exposure is better, but it does not ensure the safety. Furthermore, according to radiation hormesis described in Sect. 2.7 in Chap. 2, exposure to a certain amount of absorbed dose equivalent, a little higher than natural radiation exposure, can have a health promotion effect. There are reports that severe environments can promote growth of plants. If some kinds of stimulus are necessary to promote healthy growth, natural radiation may have some contribution as the stimulus for human beings. This is discussed again in Chap. 8.

The annual average absorbed dose equivalent caused by natural radiation in the world is around 2.4 mSv. In history, human beings have been acquiring some resistance or resilience to continuous exposure to the natural radiation, and even the exposure might be necessary. (Please remind that this statement does not mean to recommend the low-dose exposure.)

The average annual absorbed dose equivalent in Japan is about 6 mSv, of which about 1/3 of 2.1 mSv is due to the natural radiation, and about 2/3 of 3.7 mSv is given by medical purposes, which is considerably larger than the global standard. This means that the benefits of the medical exposure surpass the negative effect of the exposure. In developing countries, it is difficult to receive radiation therapy, so it is not easy for human beings to evaluate the benefit given by the medical exposure.

4.1.2   Definite and Stochastic (Probabilistic) Effects of EQ Exposure

There is no threshold in absorbed dose equivalent to ensure the safety of the EQ exposure. This is also true for exposure to dangerous chemicals, in particular, agrichemicals. For the agrichemicals there is no threshold for health damage, too, i.e., absolute safety is not insured with any small amount of their intake. Usually, the appearance of health damages caused by any medicines including agrichemical are very scattered. Hence their effects are evaluated for every 100,000 people, so as the damages caused by EQ exposure, and are referred to as the stochastic (probabilistic) effects of the medicines and the EQ exposure. For the evaluation of the chemical damage, health data for 100,000 people can be obtained, because various chemicals are widely used. In contrast, for the EQ exposure, it is not possible to conduct the exposure tests for such large number of people so that the data is only accumulated for people who have happened to be exposed to EQ by accidents, or for professionals who are intentionally exposed. The data of people exposed to the radiation from atomic bombs in Hiroshima and Nagasaki are quite valuable and have played an extremely important role in assessing the impact of the EQ exposure, and they will continue to be important even in future. The actual number of people exposed to EQ and suffered from health damage is so small that it is extremely difficult to confirm the damage was really caused by the EQ exposure even for a few people in 1000 people.

Therefore, we have collected the data of people exposed to higher absorbed doses, clearly exhibiting some health damages or diseases, for ex. people exposed at Hiroshima and Nagasaki, treated with radiation therapy, exposed by accidents, etc., and make some theory or model that shows the probability of appearance of diseases by EQ exposure. Then the extrapolation of the theory or model allows to estimate the risk for lower absorbed dose, and predict the stochastic (probabilistic) effect.

EQ exposure with the absorbed dose equivalent of over 1 Sv in a short time will certainly affect individuals. Such high absorbed dose of the EQ exposure clearly hurt

Table 4.1 Lethal dose for living beings [2, 3]

Living beings	Lethal dose (Gy)
Mammals	5–10
Insects	10–1000
Trophocyte in bacteria	500–10,000
Sporule in bacteria	10,000–50,000
Virous	10,000–200,000

anyone to give disease or apparent health disorder, which is referred to as definite effects, or non-probabilistic effects in the 1977 ICRP recommendation [1]. Hair loss, infertility, cataracts, etc. are known as the effects of the EQ exposure. The probability to get such diseases increases with the absorbed dose equivalent. The threshold over which each disorder occurs is defined experimentally as summarized in Fig. 2.2 in Chap. 2. The lethal dose for short-term exposure is also examined as shown in Table 4.1 [2, 3]. However, the lethal dose which is the value to result in the death of about a half of the living beings exposed with the same absorbed dose varies greatly over several times depending on individuals. The lethal doses in the table are given in Gy (absorbed dose), but not in the absorbed dose equivalent (Sv). As mentioned earlier, there is not much difference between Gy and Sv. When a person is exposed to several Sv (1 Sv = 1000 mSv) at one time, they are certainly at risk of their life.

The lethal dose of mammals is several orders of magnitude smaller than that of viruses and bacteria. The higher the organisms are, the more complex their tissues and organs are and the more susceptible to the EQ exposure.

4.1.3 Evaluation of the Effects of Low-Dose Exposure and Reduction of Exposure

There is no doubt that the absorbed dose should be kept as low as possible. In order to avoid the health damage by EQ exposure, an upper limit on the absorbed dose is set separately for ordinary people and professionals who are forced to be exposed to EQ or who are working in a radiation environment, such as the Fukushima nuclear power plant. The upper limit is decided considering canceration rate in daily life (For ordinal people in Japan, the annual death toll caused by cancer is about 350,000 (about 300 for 100,000 people), about 30,000 suicides (25 people), and about 5000 people (4 people) in traffic accidents). However, the canceration rate in daily life varies greatly depending on the lifestyle of individuals. For example, it is known that smokers are about three times more likely to suffer from lung cancer than non-smokers. Not only the lifestyle of the individuals but also resilience or recovering ability from disease are greatly different depending on the individuals. Therefore, the canceration due to EQ exposure varies appreciably from person to person. Considering such individual differences, International Commission on Radiation Protection (ICRP) has set the absorbed dose equivalent as about 20–100 mSv per year with which about 1 person for 1000 people will be affected. The value is extrapolation of data for people exposed to higher absorbed doses to lower absorbed dose range as shown in Fig. 4.1. The latest recommendation, which has made in 2007, is that "For public protection in an emergency, the Committee has set reference levels in the range of 20 to 100 mSv as the highest planned absorbed dose equivalent by national agencies. (ICRP 2007 Recommendation, Table 8) [4]. Based on this, Japanese government has designated areas with an estimated annual radiation dose of 20 mSv or more in the affected areas of the Fukushima nuclear power plant as an evacuation zone, and has set the regulation level for the annual absorbed dose of residents in the affected areas to be 20 mSv or less. Before the accident, the upper limit for ordinary people had been set to be 1 mSv per year, so it was raised 20 times at a time. Owing to this change the anxiety for EQ exposure seems spreading in residents in the affected area. The value of 20 mSv is based on ICRP recommendations and reasonable. However, for safety, the Japanese government had set the value lower, i.e., 1/20 of 20 mSv which was not based on the academic basis. In normal operation of nuclear reactors in Japan, discharges of radioactive material are regulated to be 1/10 from ICRP standard in consideration of safety. Such unnecessary reduction might have enhanced the scary feeling of EQ exposure.

Returning to the first question, "Is there any critical value in absorbed dose equivalent (in Sv) to distinguish secure and insecure?" The answer is "there is no critical dose". Changing the question to "Is it absolutely safe to be exposed with the exposure dose equivalent of 20 mSv or less per year?" The answer is again "No". Then you may ask "Is the exposure of less than 1 mSv secured?" Still the answer is "No". The exposure of around 100 mSv or less dose is not likely to give any immediate effects. However, the concern remains for later years. There could appear some

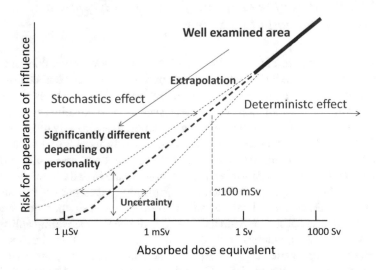

Fig. 4.1   Risk for appearance of influence of EQ exposure with absorbed dose equivalent

effects, especially on children. However, the effect, if it appeared, would be quite different depending on individuals and their living environment. It is hardly possible to mention the cause of some diseases that appeared a few decades after the EQ exposure.

In any way, concerns on the appearance of the influence of long-term exposure are larger for the internal exposure, i.e., the EQ source is in a human body. For the external exposure, γ-photons are main cause, while for the internal exposure, all kinds of EQ including α- and β-particles, and γ-photons give damage to tissues or organs ingesting EQ sources. As mentioned in Chap. 1, the EQ sources were taken into the body are exhausted according to their biological half-life. However, except for tritium, the half-live is not short, i.e., more than weeks. Therefore, the internal exposure continues until the source is exhausted with the biological half-life or artificial assistance of a medicine.

The external exposure could be reduced by removing or being away from the source. In case of the Fukushima accident, the EQ sources were widely distributed and deposited on the soil. After 10 years from the accident, although some are penetrated deep in the soil, most of the sources still remain on the ground surface, buildings, and outdoor structures, which can be removed with something like adhesive tape. (Since radioactive material adheres to the tape, it should be kept in a container to protect the exposure.) If vegetables are washed with water, a significant part of the sources on their surface can be removed.

If a γ-photon source is taken into a human body, it can be detected from outside and one can determine the intensity of the source and the absorbed dose equivalent to or effective dose of a specified organ. It should be mentioned that the absorbed air dose equivalent in nature is less than 2.4 mSv per year, which is no need to fear. However, a feeling of fear for the air exposure and resulting stress would reduce resilience and enhance radiation sensitivity.

Since α- and β-sources taken into a body cannot be detected unless the source is near the skin, we have to rely on the inspection (detection of EQ) of its excrements. Since ^{131}I, one of the most intense β sources released at the Fukushima accident is easily taken into thyroid gland, but is hard to remove, raises cancer probability. This is the one of the reasons to reduce the upper limit of absorbed dose equivalent for children to be 1 mSv y^{-1} from 20 mSv y^{-1} for adults. As mentioned several times, it is not difficult to avoid the external exposure; to stay away from or to remove the source. Even if the source adheres to surfaces of buildings and social structures, it can be removed by washing with water, etc. It is very unlikely that the source will be taken in from the skin, so there is no need to be afraid. However, very fine particles dispersed by the hydrogen explosion at the Fukushima nuclear power plant may contain a fairly high level of activity. These can be detected even if they are about 1 m away. The location of these particles on the surface of ground or structures can be detected by sequential measurements, for example, with each time at several tens of cm apart from the surface. If there is no additional fall-down of EQ sources, which can be easily detected, no need to worry about the air exposure.

The internal exposure should be avoided as less as possible. It is very unlikely that the source will be taken into the body from the skin, but if it enters with exhalation or as a food, it likely accumulates in a particular organ (e.g., the thyroid gland if it is iodine). If children played on the field and were covered with mud including the source, they could swallow it. Then it would be quite difficult to remove. If the absorbed dose equivalent in the field seems slightly higher, it is necessary to avoid ingestion or inhalation with using a face mask.

EQ is invisible but easy to detect. Using a reliable measuring device, continuous measurements will ensure safety and effectiveness to avoid unnecessary fear. In any way, one may ask "Is it unable to theoretically elucidate the effects of EQ exposure?" In the following sections, described are how the effects of the EQ exposure on substances appear in different scales in sizes and times, from nanometer (nm) scale (atomic/molecular level) to meter (m) size (human body) with separation of inorganic and organic materials, and living beings.

4.2 Effects of EQ Exposure in Inorganic- and Organic Materials, and Living Beings

In discussion of the effects of EQ exposure or irradiation, main focus is given to the effects on living beings. However, in order to understand how the effects of EQ exposure appear, basic knowledge of energy absorption (deposition) processes in substances or materials given by the EQ exposure is required. The energy absorption processes are different depending on the type of EQ as described in Chap. 2, and at the same time, the influence of the absorbed energy varies depending on characters of the exposed substances.

Among various living beings, as shown in Table 4.1, higher their order, the more complex their tissues are and the more susceptible to the EQ exposure and accordingly the lower the lethal dose. The influence of the EQ exposure in inorganic materials is less than that of organic materials and much less than the living beings. For example, metals show quite high radio resistance. A pressure vessel of a nuclear reactor made of steel can be used for more than 40 years of operation, while human beings cannot be tolerant in a reactor core even in 1 s. Of course, the pressure vessel gradually becomes brittle by neutron irradiation so that the life of the reactor is determined by the embrittlement of the pressure vessel (about 30–60 years) because nuclear fuels and other internal structures of the vessel can be replaced, while the vessel is not.

As such the influence of the EQ exposure varies appreciably depending on what kind of substance is exposed, whether it is inorganic materials, organic ones, or living beings. Even for inorganic materials which are the most tolerant to the EQ exposure, the influence is quite different depending on their chemical nature, either metals, covalent bonding materials, or ionic bonding materials.

4.2.1 Effects of EQ Exposure in Inorganic Materials

When a substance is exposed to EQ, it absorbs some or all of EQ energy. The energy absorption process is different depending on whether EQ is charged or noncharged. For charged ones, such as electrons and ions, the Coulomb interaction between the incident EQ and electrons belonging to the constituent elements of the substance causes electron excitation or ionization, i.e., electrons are released from their binding states in constituent elements (atoms). Therefore, charged EQ is called "**ionizing radiation**".

In case of noncharged EQ-like photons (electromagnetic waves) and neutrons, their initial interaction with the substance is not the Coulomb interaction. Hence, they are called as "**non-ionizing radiation**". However, energetic photons like γ-photons and X-rays collide with electrons giving some or all of their energy, which is referred as the Compton collision or effect. Once electrons are released by the Compton effect of incident photons, they can succeedingly ionize constituent atoms of the substance. Neutrons collide directly with nucleus of constituent atoms of the substance and displace them from their original positions and the displaced atoms are mostly ionized, of which cross-section is quite different among their atomic and mass numbers. Therefore, the distinction between the ionizing and non-ionizing radiations is not very important. (If the energy of photons and neutrons is too low to ionize atoms, it can be said that they are the non-ionizing radiation.)

As described above, energy absorption processes in a substance, in other words, energy loss processes of incident EQ in the substance, are divided into two different schemes, one losing energy by electron excitation (electron energy loss) and the other by nuclear collision (nuclear energy loss) and the partition between the two changes depends on the kind of EQ and their energy. Figure 4.2 shows how energy of high-energy ion/electron incident to a substance is lost in the depth with separation of two energy loss processes. When the energy of the incident one, E, is large, the ionization or electron excitation dominates the energy loss process, keeping energy loss rate or stopping power (dE/dx) nearly constant. Therefore, this energy loss process is called linear energy transfer (LET) regime. In case of electron injection, the energy of injected electros is mostly lost with LET. When the energy becomes lower, in case of ion injection, nuclear collision becomes dominant. Since the energy loss due to a single nuclear collision is much larger than the single electron excitation, the incident ion stops at short distance compared to the electron energy loss region. The energy loss rate is maximized near the end of the ion trajectory (called the projected range: Rp) which is called Bragg Peak. The vertical distance from the surface to the stop point of the incident ion corresponds to Rp. Since the incident ion penetrates in the substance repeating collisions, its actual traveling distance is much longer than Rp.

Once the type of EQ and their initial energy are known (determined), their project range in a human body can be estimated. This is applied to proton therapy, or heavy particle therapy for cancer treatment, that is, protons or heavy ions are injected to

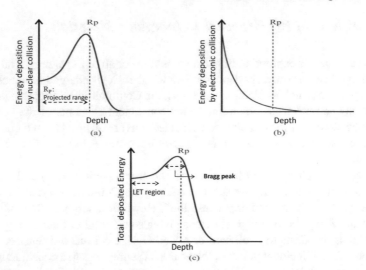

Fig. 4.2 Deposited (absorbed) energy profiles in depth under energetic ion injection. Those of nuclear collision (**a**), electron excitation (**b**), and the total (**c**) are separately given. R_p is the projected range, surface normal trajectory of the incident ion

kill cancer tissue adjusting their Rp to be the same depth of the location of the cancer tissues (see Sect.7.3 in Chap. 7).

Here one should note that the LET regime dominates in the energy loss process of high-energy β-particles or γ-photons, and accordingly, the radiation weighting factor for both is the same as given in Table 2.1 in Chap. 2.

Figure 4.3 schematically shows main interactions with ionizing radiation (EQ) of α- and β-particles and γ-photons, and neutrons in a substance. For α-particles, the ionization and electron excitation, and the generation of Bremsstrahlung caused by changes in the trajectory of ejected electrons dominate their energy loss, while for β-particles the Coulomb interaction, and for γ-photons Compton process and following Coulomb interaction. These energy loss processes correspond to Fig. 4.2b. α-particles or heavy ions appreciably lose their energy near Rp colliding with the constituent atoms to be displaced as given in Fig. 4.2c. Neutrons cause nuclear reactions in addition to the atomic displacements. Since the displaced atoms and transmuted atoms produced by the nuclear reactions are ionized, the subsequent energy loss process will be the same as that occurs for ions like an α-particle. From these points of view, it does not make much sense to distinguish the ionizing- and non-ionizing radiations.

The energy loss of γ-photons by the Compton process is reflected in the energy spectrum of γ-photons emitted from ^{137}Cs given in Fig. 4.4. As shown in the figure, ^{137}Cs emits γ-photons of 662 keV as detected the highest peak. In addition, broad emission below 662 keV is detected, which indicates some of the energy of 662 keV is lost to ionize inner shell electrons of Cs atoms. The Compton edge in the figure appears at 254 keV lower than 662 keV which is the binding energy of 4 s electron

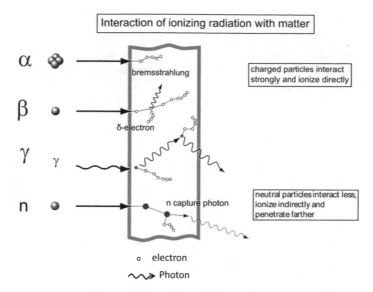

Fig. 4.3 Phenomena induced by the injection of EQ in a substance. Dominant phenomena are ionization, electron excitation, Bremsstrahlung emission, and Compton scattering. They are separately given for α- and β-particles, γ-photons, and neutron. Those given by ionization radiation are mostly ionization of atoms in the substance caused by Column force. Although the initial collision of non-ionization radiation is not influenced by the Column force, the recoil atoms are ionized. Accordingly, the secondary and after collisions are controlled by the Coulomb force. (See also Fig. 1.8 in Chap. 1)

Fig. 4.4 Energy spectrum of γ-photons emitted at disintegration of ^{137}Cs observed with a semi-conductor spectrometer

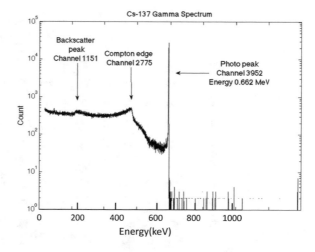

of Cs. Repeating the Compton scattering with electrons of 662 keV further loses the
energy as appeared broad spectrum to lower energy region.

In the following, the effects of EQ exposure that appeared in substances are
explained separately for metals, covalent materials, ionic bonding materials, organic
materials, and living beings.

4.2.1.1 Effects of EQ Exposure in Metals

In metals, constituent atoms are regularly arrayed to make a lattice, as shown in
Fig. 4.5. When EQ is injected into a metal, they lose energy by electron excitations
and nuclear collisions until they fully lose their energy and stop near the projected
range.

Damages Caused by Nuclear Collisions

In nuclear collisions, the incident EQ collides directly with the nucleus of a lattice
atom to displace it to be an interstitial atom and remain a vacancy, referred to as the
formation of a Frenkel pair as shown in Fig. 4.5. In other words, the energy brought
by EQ is used to move the position of the lattice atoms (displacement damage),
which requires the energy of around 50 eV and above. Although numbers of the

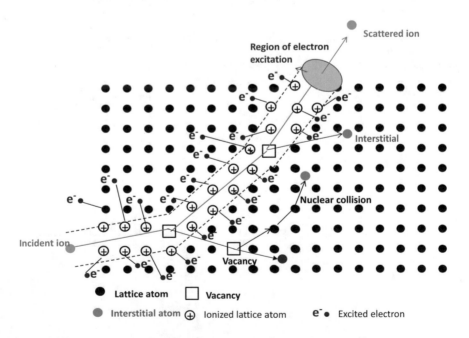

Fig. 4.5 Schematic drawing for energy loss processes of energetic ion incident to a metal

Frenkel pairs are generated, most of them disappear with the recombination of a vacancy and an interstitial atom. Consequently, only a few vacancies and interstitial atoms remain long. Under the EQ exposure, some interstitial atoms come together to be interstitial clusters and loops, and some vacancies come together to be vacancy clusters and loops. Once these clusters or loops are formed, they can remain in the lattice and changes physical properties of the metal. Consequently, the irradiated metal becomes hard and loses ductility (becomes brittle). After heavy irradiation, the metal may crack like the glass breaking. This is the most important influence of the EQ exposure of metals referred to as loss of ductility or irradiation embrittlement. The lifetime of an iron-based reactor pressure vessel of a nuclear reactor is limited by the loss of ductility caused by neutron irradiation. This irradiation damage caused by atomic displacement becomes severer for heavier EQ, while the damaged depth in the irradiated metal becomes shallower. This is why the heavier materials are more difficult for the incident EQ to penetrate deeper as described in Fig. 1.1 in Chap. 1 and accordingly work as a radiation barrier. Neutrons are light but have no charge so that they are easy to approach the nuclei and often cause nuclear reactions, nuclear transmutation, and fission of nucleus as mentioned in Chap. 2. The nuclear reaction releases high-energy as EQ including ions of transmuted nuclei, which in turn gives additional radiation damage. Some of transmuted nuclei are radioactive, i.e., the transmuted nuclei store excess energy and radiate the excess energy as EQ (either, α-, β-particles, and/or γ-photons).

The embrittlement of the reactor pressure vessel is inspected by examining the degradation of mechanical properties of test pieces placed in the reactor made of the same material used as the vessel with certain time interval. If the degradation of the mechanical property including the ductility loss is well within designed values to keep healthiness of the pressure vessel, the pressure vessel can be safely used at least until the next inspection.

Damage Caused by Electron Excitation

Any charged EQ or ionizing radiation incident in metals causes electron excitation to make excited electrons and ionized atoms. Figure 4.5 schematically shows energy loss processes of an incident ion consisting of the electron excitation process and nuclear collision producing Frenkel pairs. The energy of several eV is required to ionize atoms, which is much smaller than the energy required to make atomic displacement mentioned in the previous section. Therefore, when charged EQ are injected into a substance, they excite a lot of electrons along their trajectories of remaining ionized atoms, as indicated in Fig. 4.5. Since there are many freely moving electrons (conduction electrons or free electrons) in metals, excited electrons by EQ collide immediately with the freely moving electrons giving their energy, and are de-excited (relaxed). Accordingly, the effect of the electron excitation introduced in metals by EQ exposure is not so appreciable compared to the displacement damage. On the other hand, the electron excitation plays a major role in the covalent bonding materials, molecular crystals, and organic substances as described in the following sections. It should be mentioned that in organic materials electron excitation often breaks bonding between constituent atoms. Accordingly, living beings are quite sensitive to EQ exposure. Because electrons are much lighter than atoms, the required energy for electron excitation in organic materials is smaller. Therefore, the organic materials and the living beings are more likely affected by excited electrons and the effects of the EQ exposure are noticeable at much lower absorbed doses than in metals. Since the most of damage caused by the electron excitation appears in LET region in Fig. 4.2, those EQ that dominates electron energy loss in their energy loss processes is referred to as high LET radiation.

4.2.1.2   Effects of EQ Exposure in Covalent and Ionic Bonding Materials

In covalent and ionic bonding materials, different from metals, valance electrons are localized between neighboring atoms bonded together. When the valence electrons are exposed to EQ and get an energy of a few eV (different depending on the bond strength), they are excited or ionized to break the bond resulting the decomposition of the compound. This is called radiolysis of covalent and ionic bonding materials. As with metals, most of excited electrons are relaxed (de-excited) or recombined to the ionized atoms and return to their original state, and only few broken bonds remain.

Water radiolysis is a well-known process as schematically shown in Fig. 4.6 [5]. First, EQ gives energy to electrons in H_2O molecules resulting in electron excitation or ionization occurs. Then the excited electrons collide with other H_2O molecules to produce various ions and radicals such as OH*, OH^-, H*, H^+, O*, O^-, O_2^-, H_2,

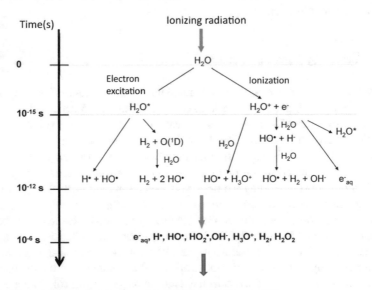

Fig. 4.6 Water radiolysis, products, and their sequential changes [5]

O_2, H_2O_2, H^+, e^-, etc., which have a life of 10^{-6} s or a little longer. ("*" indicates a radical or excited state.) Since valence electrons contributing to an O–H bonding are usually paired, H_2O molecules are written as H:O:H in high school textbooks with ":" showing two electrons. Since there is only one electron in H, all electrons in H_2 contribute to the bonding. On the other hand, among eight electrons of O only two contribute to the O–H bonding. H* and OH* is produced when H_2O decomposes with one valence electron is shared with each other. They are referred to as radicals (active atoms or active molecules) and are unstable so that most of them return to H_2O, but some turn to be H_2, O_2, H_2O_2, etc. They could react with some other molecules in water if they existed resulting in different molecules. These processes are schematically shown in Fig. 4.6.

In radiation chemistry, an important parameter named as the **G value** is introduced to show how many water molecules are decomposed when the energy of 100 eV is given to or absorbed in 1 g of a substance by EQ exposure. The G value for water irradiation is determined to be 8. Since the energy of 5.1 eV is required for the decomposition of a water molecule, only 2/5 of deposited energy of 100 eV is used (5.1×8 eV/100 eV $= 40.8$ eV) for the decomposition. Even though more than seven water molecules could be once decomposed, they were recombined to water (recovering).

When different molecules are present in water, they could directly react with the radicals immediately after the water decomposition or a little later with the radicals having a slightly longer life. Since, in any living beings, water dominates in their cells, the water radiolysis is the main cause of the appearance of biological effects through the reactions of radicals with genes, DNA, and RNA. This will be explained in more detail in the next section.

In Fig. 4.7 [6] are summarized such processes caused by EQ exposure in living beings from initial injection of EQ occurring within 10^{-15} sec to final damage appearance, days, and years after. Compared with Fig. 1.2 in Chap. 1, Fig. 4.7 includes the resilience or repairing ability of the living beings, which is discussed in Sec. 4.3.

Different from water, in crystals or glass of covalent compounds the constituent elements (atoms) are located at the fixed positions or at the lattice points. Therefore, atomic displacement is induced by EQ exposure similar to metals. Since covalent bonding materials are usually consisting of different elements, the effect of displacement is somewhat different from metals. For example, in a crystal or glass of silica (SiO_2), oxygen is lighter, so that oxygen is easier to be displaced from the lattice position leaving an oxygen vacancy. In SiO_2, O is negatively charged, while Si is positive and the oxygen vacancy can trap an electron. The electron trapped at the oxygen vacancy thus produced is referred to as a color center, because the trapped electron at the oxygen vacancy can be excited by violet light and relaxed releasing visible light (looks like colored). Using this principle, radiation detectors called a glass batch or glass dosimeter are constructed. Alternatively, displaced Si atoms can accumulate to be Si clusters that absorb visible light. Therefore, when SiO_2 is exposed to γ-photons, it becomes brown colored as shown in Fig. 4.8.

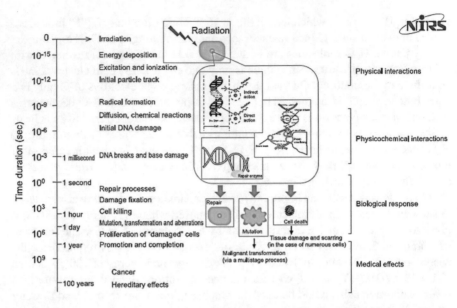

Fig. 4.7   Processes caused by EQ exposure in living beings (in [6], Open for public)

Fig. 4.8   Color change of a glass bottle by γ-photon exposure

4.2.2   Effects of EQ Exposure in Organic Materials

The effects of EQ exposure on organic materials are basically caused by the bond breaks of C–C, C–H, C–O, C–N, P–H, etc. in organic molecules through the excitation of the valence electrons contributing to their covalent bonds and hydrogen bonds (like water radiolysis). If there are other molecules around the broken bonds like water, it may react with the broken bond to make different bonds from the original ones. This is used for bridging of polymers to enhance their hardness which can be caused by either β-particles (electrons) or γ-photons. Figure 4.9 [7] schematically shows the bridging of polymers by an electron beam. The injected electrons break C:H bonds

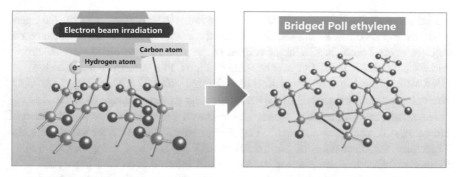

Fig. 4.9 Radiation-induced bridging in polymers (in [7], Reprinted with permission)

in the polymer chain and release two H*, which subsequentially recombine to H_2 leaving dangling bonds on C atoms in the chain. Then two neighboring dangling bonds combine to make a new C–C bond instead of C-H bonds resulting in the bridging of two chains. This technique with the electron beam is widely used in industries to improve properties of polymers, such as heat resistance, hardening, electric resistance, and so on because electron beam is easier to handle compared to γ-photons.

Basically, the bridging increases with the absorbed energy or dose resulting in hardness enhancement. However, the higher dose exposure results in the cutting of the molecular chains to be shorter and the polymers becoming brittle like weathering. Weathering of the polymers in nature is the bond breaking by oxygen (oxidation) similar to that caused by EQ exposure.

As such, the influence of EQ exposure appears as a chemical reaction in materials. Particularly important is that the EQ exposure can break bonds which is hard to be broken in normal chemical conditions. If this happened on DNA or RNA in a cell, malformation of the cell or mutation of the tissue containing the cell could be caused, as discussed in the following section. Of course, certain chemical agent makes it possible to lead to chemical reactions which do not occur in nature. For example, some special chemicals such as thalidomide and dioxins damage DNA, like the EQ exposure.

4.2.3 Effects of EQ Exposure in Living Beings—From Molecules in Cells, Tissues to Individuals

The effects of EQ exposure in living beings first appear as intracellular damages. In cells, of course, there are two damaging processes, as discussed above, i.e., displacement of atoms and electron excitation. The displacement causes a direct break with the collision of a constituent atom of DNA with the incident EQ. The electron excitation causes both direct and indirect bond breaking, the former is caused by electron excitation of atoms belonging to DNA and the latter in the cell near DNA. Damages given by the electron excitation far exceed the displacement effects. Because the electron excitation occurs in larger volume of the cell, and generated ions, electrons, and radicals attack DNA to give chemical changes, like substitution of an atoms of DNA with a foreign atom, insertion of a foreign atoms, or breaking of the DNA chain. In this respect, special drags can also cause similar damage to DNA.

As mentioned in previous section, the electron excitation is the major cause of the damage of inorganic materials. For living beings, it gives more serious effects. Therefore, the discussion of the effects of EQ exposure of living beings mainly focused on effects of the electron excitation and referred to as LET damages.

Figure 4.10 [8] schematically shows how the electron excitation in EQ exposure damages DNA, while Fig. 4.11 [9] shows what kind of chemical reactions is caused by the EQ exposure. There appear direct bond breaks of DNA either by the displacement

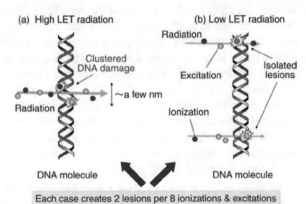

Fig. 4.10 Damaging of DNA caused by γ-photons exposure. Difference in absorbed dose rate is schematically compared as **a** high LET and **b** Low LET radiations (in [8], Reprinted with permission)

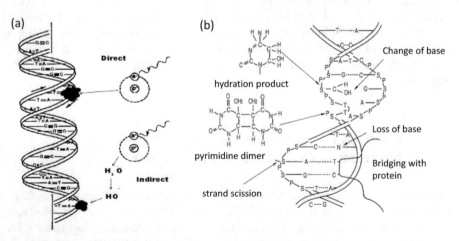

Fig. 4.11 **a** Direct and indirect attack of DNA by H⁺ or OH produced by water radiolysis, **b** various damages appeared in DNA attacked (in [9], Reprinted with Permission)

of atoms or the excitation of electrons belonging to DNA, and indirect breaks or chemical changes by the electron excitations producing electrons, ions, and radicals in its surrounding area (referred to as radiolysis). During EQ moving in cells and tissues of the living beings, they continue to excite electrons along their trajectories to make radiolysis. The influenced volume by the radiolysis is within the radius of around 1 nm along their trajectory of the incident EQ. The trajectory significantly changes with the energy, charge state, and mass of EQ. In Fig. 4.10 are separately shown two cases: (a) High LET case, i.e., the number of excited electrons is larger, and (b) Low LET case, i.e., the number of excited electrons is fewer.

Since the size of EQ is quite small, the probability of the direct break of DNA is much smaller than the indirect breaks caused by the radiolysis of water surrounding DNA. As mentioned earlier, the water radiolysis produces various radicals, which have rather longer life and a very strong oxidizing power. When some chemical reactions to change or break of bonding in DNA occurs in cells, subsequently, significant damage is caused in tissues/organs and a body. In addition, since the cell contains a certain amount of oxygen molecules (two to three percent in normal fibroblasts), electrons released by the ionization effect of EQ exposure generate O_2^- ions, which is one of the active oxygen species hazardous to health. H_2O_2 produced by water radiolysis plays similar role as the active oxygen species. Thus, the EQ exposure generates the active oxygen species in cells. Whatever the cause of the active oxygen generation, either EQ exposure, chemicals, or stress, their influence on the cell and organs is the same.

On the other hand, there are antioxidants in cells that pretend the action of the active oxygen species. As an example, the activity of O_2^- ion is suppressed by a chemical agent called Super Oxide Dismutase (SOD), and that of H_2O_2 by an enzyme called catalase. A substance called glutathione also works as the antioxidant. Although cells are always exposed to active oxygen species by oxygen respiration, such antioxidants are daily working in them. Similarly, some antioxidant systems might have been developed in living beings for the EQ exposure over long history, which would give some resilience to low-dose EQ exposure on the human body. However, the resilience is not quantitatively evaluated which is one of the causes for large uncertainty of the appearance of effects of the low absorbed dose EQ exposure. As is well known, the resilience also varies greatly from person to person and depends on individual's environment and mental state. If one lives in healthy condition for both mind and body, he can continue to be healthy, and is likely to be able to reduce the influence of the EQ exposure as well.

It should be mentioned that the resilience is not limited to cells, but to tissues and organs. A body has also some resistance to the EQ exposure. Normally, when DNA in a cell is damaged, the cell loses ability to regenerate the cell resulting in its death. However, because the tissues have the function of eliminating dead cells, the death of the small number of cells does not necessarily lead to cancer or the death of the tissue.

One cannot reduce the absorbed dose after EQ exposure. However, if he believes the resilience, he might reduce symptoms or damages caused by the EQ exposure. It would be a good way to live avoiding despairing of the exposure and believing the resilience.

4.3 Resilience to EQ Exposure and Recovery

In previous sections, described is how EQ exposure causes damage to materials. At the same time, the materials have the ability to recover somehow the generated damages. Accordingly, damages that remain after the exposure are only a small part of the initially generated ones. Even the damages that remained after the exposure could be also recovered over time, or sometimes could get worse.

The damaging processes in inorganic materials (metals, covalent, and ion-binding materials are collectively referred to as inorganic materials) have been fairly elucidated theoretically. The amount of the defects initially generated has become almost quantitatively predictable. However, changes in the initial defects to defect clusters or formation of damages are difficult to predict. Furthermore, the recovery of the initial defects or damages simultaneously occurring with the damage processes is not understood well. Only the observation of the irradiated material afterward gives information on how much damage remained and how damage structure changed with the assistance of recovering ability. It should be noted that the formation of the initial defects does not depend much on the temperature of the materials, while the recovery processes vary greatly depending on the temperature of the material. As a matter of fact, most of the initially generated recover or disappear quickly, and only a small number of damages is left. Therefore, it can be said that the effects of EQ exposure are rather determined by recovering ability or resilience.

It should be noted that the effect of EQ exposure is hardly seen in liquid metals. Since the constituent atoms of the liquid are always moving, the displacement of the atoms does not matter and a vacancy is immediately filled by an atom nearby.

When the displacement occurs near-surface, displaced atoms can escape from the surface referred to as sputtering or enhanced sublimation. Those phenomena are not noticeable unless the absorbed dose rate near the surface is very high.

In covalent and ionic bonding materials, the binding energy between the neighboring atoms is generally greater than that in metals. Therefore, the number of displaced atoms by EQ exposure is less compared to metals. However, the difference is not so large, because the incident energy of the EQ is very high. On the other hand, because the covalent and ionic bonding materials are made up with different kind of atoms, for example, oxides made up of oxygen ions and metal ions, the same kind of displaced atoms can gather to form clusters or change bonding character leading to coloring or blacking by the EQ exposure as shown in Fig. 4.8. In atomic collisions, the heavier the constituent atoms, less the number of displaced atoms, while the larger the energy loss at one collision results in larger energy deposition in a unit volume or smaller volume affected.

Different from inorganic materials, organic materials are consisted of larger numbers of different kinds of atoms and show complex structures in atomic and molecular arrangements. Cells in tissues consist of various kinds of molecules and compounds. In particular, DNA, the most important in a cell, is surrounded by water molecules and other molecules that are completely different from DNA. Therefore, the break of some bonding in DNA could introduce different arrangements in DAN caused by new bonding at the broken bond. Once the different arrangement is introduced in DNA, the recovery becomes hardly possible. Since binding energy between atoms in organic materials is generally lower than that in the inorganic materials, bond breaking is easier than that in the latter. Since in cells of living beings, the damage in their DNA or RNA can lead to their death or canceration of tissues, the effects of EQ exposure easily appear in tissues and in a body. Of course, the effects appear differently depending on what kind of tissues or organs are exposed to EQ.

In summary, the higher the organism, the more different types of molecules it is consisted of, and the more complex the structure of the molecules is. This makes recovery more difficult, i.e., radiation sensitivity increases. In other words, the higher the living beings as shown in Table 4.1, the lower is their lethal dose. Moreover, the appearance of the influence of EQ exposure becomes complicated for higher living beings and difficult to predict. Accordingly, human beings are most susceptible to the influence of EQ exposure compared to other living beings, and the appearance of the exposure effects is the most complex. At the same time, the difference in resilience by individuals is large.

In this way, for human beings, the appearance of the effects of lower absorbed dose EQ exposure are quite diverse, and quantitative evaluation of the effects of the EQ exposure requires a large amount of data. The lower the absorbed dose, the less is its impact so that more data are required for the assessment of the impact of the EQ exposure. Consequently, it is more difficult to ensure the accuracy of the assessment. Since it is not possible to examine the effect of EQ exposure with human bodies, the observation of the effects of EQ exposure has been extensively done with using microorganisms, mice, rats, and large animals. However, in such living beings compared to human beings, as seen in Table 4.1, a much higher dose is required to appear exposure effects. (There is data for EQ exposure with such a high dose of 10^5 Gy for viruses). The effects of EQ exposure to the tissues, organs, and body of human beings are predicted by the extrapolation of these data to lower absorbed dose as indicated in Fig. 4.1. Therefore, the accuracy of the prediction from the extrapolation becomes very poor.

4.4 Volume Influenced by EQ Exposure and Absorbed Dose (Deposited Energy)

In previous sections are described how EQ energy is absorbed in materials exposed to EQ and the appearance of effects of EQ exposure. Here, described is the importance of area and volume of the human body on the appearance of effects of the EQ exposure. The main point is that the influence of the EQ exposure initially appears in very tiny area or volume. The appearance is significantly different depending on where the area or volume exposed to EQ locates in tissues, organs, or a body, on the kinds, intensity, and energy of EQ, and on geometrical relation of the exposed location and the EQ source.

It is often misunderstood that the EQ exposure means all surfaces and volume of a body is uniformly exposed to "radiation" which is invisible and this makes people fear the radiation or believe the radiation being scary. For exposure to natural radiation this may be true because, in nature, radiation sources are distributed almost uniformly in space and on the surface. However, exposure over the natural exposure dose is caused by EQ sources separately from the natural radiation. The EQ sources are not necessarily uniformly distributed in surrounding but localized as point, plane, or volume sources with limited sizes. Since the source is mostly radioactive isotopes (RI) of which nucleus decay with a specified half-life, the intensity or number of emitted EQ is given by the disintegration of RI in unit time (dpm or dps, disintegration per minute and second, respectively). A human body is exposed to the flux of EQ, which is represented as the number of EQ injecting to unit area (1 m^2) and in unit time (1 s). Hence the effect of EQ exposure varies depending on where and how large the source is, from which direction EQ comes, and on which area of the body is exposed as described in Chap. 2. More importantly, since EQ are not fluid, the body is not uniformly exposed to EQ. The influenced volume by EQ exposure or the area EQ injecting is very tiny. In other words, the total or integrated area and/or volume in a human body where energy of EQ is absorbed (deposited) is quite small.

In the following, described is how energy of EQ is deposited to or absorbed in a human body under EQ exposure. In the EQ exposure, concerned is the number of EQ and absorbed energy given by them. Suppose a person weighing 60 kg is exposed to 10^4 cps of EQ having energy of 1 MeV for each quantum. (The exposure to 10^4 of cps is about 1000 times larger than that given by natural radiation.) If all EQ energy is uniformly absorbed in a human body, averaged energy absorbed in the whole body is given as following.

The total deposited energy by this exposure is

$$1.6 \times 10^{-19} (\text{J} \cdot \text{eV}^{-1}) \times 10^6 (\text{eV}) \times 10^4 (\text{cps}) = 1.6 \times 10^{-9} (\text{J}) = 1.6 (\text{nJ}). \quad (4.1)$$

Averaged absorbed energy in unit weight is,

$$1.6 \times 10^{-9} (\text{J}) \div 60 \, \text{kg} = 2.6 \times 10^{-11} \, \text{J} \cdot \text{kg}^{-1}. \quad (4.2)$$

Since 1 Gy is 1 J/kg, the absorbed dose becomes 0.026 nGy per second. For 1 h exposure, it becomes 93.6 nGy. The total energy deposited is 1.6 nJ in 1 s or the power of 1.6 nW. Compared with an electric heater of which power is around several hundred W, the power given by EQ exposure is quite tiny with the difference of 9 orders of magnitude. Even so, this EQ exposure gives certain influence on the body. Thus, although the deposited energy by the EQ exposure is quite small, the effects on the human body appear. This would give the impression that EQ exposure (radiation) is scary.

The reason why the EQ exposure gives large influence is in the smallness of the volume or area of energy absorbed (deposited). The energy absorption given by the EQ exposure is not uniform as described above. Figure 4.12 schematically draws area and volume where energy is absorbed for exposures of α- and β-particles and γ-photons having the energy of 1 MeV with 1 cpm cm^{-2}, which is roughly equivalent to absorbed dose of 1 mGy y^{-1} for a person. Although the size of each energetic quantum is less than 1 nm, its energy is assumed to be absorbed in a cone or cylinder within 10 nm φ as shown in the figure. Then deposited power flux in circled area becomes around 10^3 W/m^2 because 1 MeV s^{-1} = 1.6 × 10^{-13} W is divided by the area of the circle (8 × 10^{-17} m^2). Compared to the power emitted from an infrared heater of around 10^3 W m^{-2}, EQ exposure gives similar power but within a quite small area of 8 × 10^{-17} m^2. There are also large differences in the energy deposited depth of EQ. γ-photons penetrate deep more than 1 m, while β-particles and α-particles within 10 μm and 0.1 μm, respectively. Therefore, the deposited energies in an unit volume are 2 × 10^4 W m^{-3}, 2 × 10^9 W m^{-3}, and 2 × 10^{11} W m^{-3}, with deposited volumes

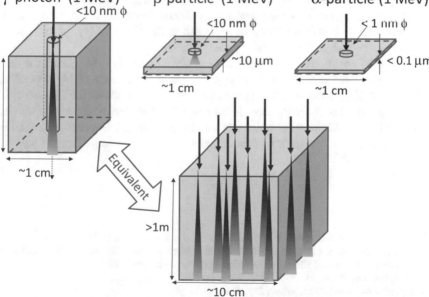

Fig. 4.12 Schematic drawing on area and volume where energy is absorbed (deposited) for EQ exposures. As EQ is selected α- and β-particles, and γ-photons have the energy of 1 MeV with 1 cpm cm^{-2}, which give nearly equivalent to absorbed dose of 1 mGy y^{-1}. Effects of the exposure are quite different depending on the location EQ hit. The overlap of the locations EQ hitting is quite seldom for low absorbed dose equivalent

of 8×10^{-17} m^{-3}, 8×10^{-22} m^{-3}, and 8×10^{-24} m^{-3}, respectively for γ-photons β-particles and α-particles.

If the EQ exposure deposits its energy on a human body (a body absorbs its energy) uniformly as a whole, the average absorbed energy density (normalized with the total weight (J kg^{-1}) or total volume (J m^{-3}) is too small to give the influence on the body. Nevertheless, quite large power is given to a very tiny area resulting the destruction of the cell which in turn could give some damage to tissues, organs, and a body. In reality, even if a person is exposed to EQ with 10^4 cps m^{-2} which is approximately equivalent of 10 μSv s^{-1} in the air dose rate, the effect will appear in a very small area of $10^4 \times 8 \times 10^{-17} = 8 \times 10^{-13}$ m^2, i.e., 0.8 μm^2. As seen in the bottom in Fig. 4.11, the influenced zones of the cones or cylinders by the EQ exposure are mostly independent and hardly overlap. Therefore, unless a human body is exposed to a dose of 10^5 times larger than the above estimation, that is, about 1 Sv, appreciable damages will hardly appear.

In addition, in a human body, there is a very large difference in radiation sensitivities among tissues and organs, and even in cells. Consequently, the influence of EQ exposure is different depending on where or which part of the body EQ is injected. Although most EQ would pass through water part in the cell, DNA and RNA could be occasionally exposed resulting in the significant damage. As such, for a low-dose exposure, even if the absorbed dose was the same, the influence could be significantly different depending on where EQ was injected. Thus, in nature, the influence of the low-dose exposure is statistically scattered and the effects of the EQ exposure appear essentially as probabilistically or statistically different.

It should be mentioned that the above calculation is rough estimation for easy understanding. Estimated absorbed dose or dose rate varies appreciably depending on the type of EQ, their energy, and substance being exposed so that the inaccuracy of more than one digit could be included in the estimated dose. Hence there is no wonder that the estimation of the appearance of the radiation effects at lower dose region in Fig. 4.1 includes the uncertainty of more than two or three orders of magnitude.

References

1. ICRP, *Recommendations of the ICRP. ICRP Publication 26*. Ann. ICRP 1(3) (1977)
2. United Nations Scientific Committee on the Effects of Atomic Radiation, UNSCERA1966 report, Scientific Annex; Effects of radiation on the environment (1966).
3. A.H. Sparrow et al., Radiat. Res. **32**, 915–945 (1967)
4. ICRP ref: 4847-5603-4313, March 21, 2011
5. S. Le Ca'r, *Water radiolysis: influence of oxide surfaces on H$_2$ production under ionizing radiation*. Water **3**(1), 235–253 (2011). https://doi.org/10.3390/w3010235
6. https://www.qst.go.jp/uploaded/attachment/1568.pdf
7. http://www.kbeam.co.jp/service/kaisitu.html
8. https://rdreview.jaea.go.jp/review_en/2007/e2007_6_5.html
9. N. Egami, *UP Biology Creatures and Radiation* (The University of Tokyo Press, New edition, July 2013). ISBN978-4-13-006504-7 (in Japanese)

Chapter 5
Reduction of Exposure, Contamination, and Decontamination

Abstract This chapter focuses on the reduction of the undesirable effect of EQ (radiation) exposure. To do this it is necessary to understand what are the results of the EQ exposure. The best way is to avoid the exposure by keeping the distance from the source or shielding the EQ. Once exposed to EQ, absorbed dose (Gy) or absorbed dose equivalent (Sv) cannot be reduced. Therefore, the mitigation of the effects of EQ exposure is for the future but not for the past. If any health hazard was caused by the EQ exposure, there is no other way than medical treatments. If the annual accumulated dose for a person is less than around 100 μSv, its risk to give health hazards is quite low, and no need to be treated.

Keywords Absorbed dose · EQ exposure · External exposure · Internal exposure · Recovery · Resilience

5.1 Introduction

"Reduction of radiation exposure", this is what everyone wants. To do this it is necessary to understand what is the EQ (radiation) exposure (in this book, we refer to the radiation as energetic quanta (EQ)) as explained in Chap. 1). In addition, it should be reminded that once exposed to EQ, absorbed dose (Gy) or absorbed dose equivalent (Sv) cannot be reduced. Therefore, the mitigation of the effects of EQ exposure is for future but not for past. If any health hazard was caused by the EQ exposure, there is no other way than medical treatments. If the annual accumulated dose for a person is less than around 100 μSv, its risk to give health hazards is quite low and no need to be treated. If any health hazard was expected, one should take some action to prevent or reduce the exposure effects.

The "reduction of exposure" discussed here does not mean to reduce the absorbed dose so far, but to avoid exposure in the future, or to reduce the internal exposure if the source is taken into the body, unfortunately. To do this, the only way is to stay away from or remove EQ sources.

In Sect. 5.2, considering the EQ exposure in the occasion of the Fukushima nuclear power plant accident, described are the possible sources of EQ for public exposure and how they were distributed, and how to remove them. This would make the readers know how to avoid the exposure. Section 5.3, are discussed internal exposure and external exposure of a human body. For the external exposure, which means the exposure is caused by EQ sources outside the body, the exposure reduction can be done to keep away from or to remove the sources. This is also true to avoid the internal exposure which is given by EQ sources taken into the body. If the sources were taken as foods or ingested as suspended matters in air into the body, they should be exhausted or discharged by some medical treatments. Isotopic replacement is an important way to remove EQ sources ingested into a body. i.e., if a radioisotope (RI) included in the source was identified, some foods or chemicals including its stable isotope could replace or dilute it. It is well known that ^{131}I is preferentially taken into thyroid. To remove ^{131}I in thyroid, recommended is to take an iodine tablet, which replaces ^{131}I with its stable isotope of ^{127}I to enhance the removal. The reduction of the internal exposure is discussed in Sect. 5.4.

5.2   Distribution of EQ Sources and Their Removal

As mentioned in Chap. 1, radiation or EQ emission is originated from their sources. Major EQ are charged particles (referred to as α– and β–particles), non-charged particles (neutron), and electromagnetic waves or light (γ–photos and X-ray). As shown in Chap. 3, there are various sources that are emitting either one, two, three, or all of them. In case of the Fukushima nuclear power plant accident, ten years ago, the sources were consisted of radioactive fission products (FPs) generated by nuclear fission reactions and widely dispersed by hydrogen explosions and wind. Accordingly, the EQ sources were widely distributed in air and on the ground surface near the Fukushima site. In a nuclear reactor, various kinds of elements as FPs are generated. Most of them are radioactive as shown in Table 3.3 in Chap. 3. Although some are chemically hazardous, like Cd (cadmium) and Te (tellurium), their amounts are too small to give health influence chemically. Among various FPs, those of larger yield or high vapor pressure, such as ^{132}Te, ^{131}I, ^{137}Cs, and ^{90}Sr are dominant to give radio hazards.

The radioactive FPs are dispersed in wild fields near the Fukushima power plant in various forms physically and chemically. However, their chemical forms have not been investigated well. Although their existence can be easily detected by radioactivity measurements, their masses are too little to identify their chemical form or to find what kind of elements are combined with them. For example, Cs which is an alkaline metal, can combine with halogens to make a salt, like CsCl, or be dissolved in water as Cs^+ and carried long distance. In fields near the Fukushima nuclear power plant, ^{137}Cs are often found in deposited particles which is insoluble in water. The particles also include materials used in the reactors as structure materials, heat insulation, and so on, and exhibit amorphous like structure.

In surrounding area of the Fukushima power plant, various sizes of particles including radioactive FPs were found with their sizes distributed from under nm to over mm and widely distributed as air suspended materials, deposited on filed, surfaces of structures, and in water in surrounding area. At first, they were blown away from the failed Fukushima plants by the hydrogen explosion, further dispersed by the wind, and deposited on the ground surface. Afterward, some were transported with rainwater, etc. Very high EQ intensity (radioactivity) is often found in localized areas for example in drain pipes on which water droplets from roofs are aggregated (concentrated) and dried out remaining radioactive ingredients.

Figure 5.1 shows how radioactive materials were deposited on vegetable leaves cultivated at the time of the accident. Red points in the figure show that the intensity of EQ (or radioactivity) was high. EQ was detected even from the backside of the leaf. The less intensity observed on the backside indicates that EQ passed through the leaf with some losses in the number of EQ and their energy. If the sources were taken inside the leaf, the intensity profile should reflect its characteristic structure, i.e., the veins of the leaf. The intensity distribution in Fig. 5.1 is clearly different from those in Fig. 1.7 in Chap. 1 which is the profiles ^{40}K included in tissues of root vegetables and reflects their structures. Furthermore, washing the leaves with hot water decreased the activity to 1/3. This also supports the activity is located on the surface. These observations indicate that the source was dominated with β–particle emission and deposited on the leaf surface. And washing works as a method for decontamination. Actuary, the source was found to be dominated with ^{131}Cs.

Again, EQ sources dispersed in the air (FP in the case of Fukushima) were differently distributed/deposited depending on their sizes and chemical nature. Very small ones floating in air and larger ones deposited on the field with rain are almost homogeneously distributed. Some showing high activity is transported and aggregated afterward by water. On the other hand, the larger ones immediately fell down

Fig. 5.1 Distribution of radioactive materials falling on a vegetable leaf released at the nuclear accident of Fukushima power plants together with the photograph. Compared are those observed on front and back sides and before and after rinsing with water. [Provided by Prof. Yoshida, Tohoku Univ., Reprinted with permission]

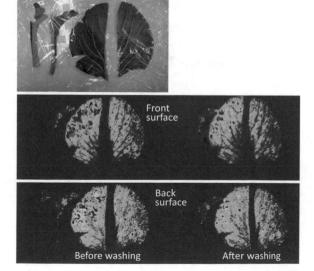

after the hydrogen explosions and were found locally as hot spots of which locations were different depending on their sizes and air condition when they had been
released. Thus, radiation sources are separated into two categories in their distribution, uniformly distributed ones in air, surface, and water, and localized ones mostly
exist as the hot spots on the surfaces of fields and structures.

If there is a hot spot caused by a localized EQ source, the EQ intensity in
surrounding area becomes high. The areal sizes/volumes showing high EQ intensity are largely different depending on the type EQ emitted from the source from
around cm^3 for β–particles to more than m^3 for γ–photons. If high EQ intensity was
detected in some local areas, there should be the EQ sources consisting of radioactive
FP falling and aggregated.

Since a radiation detector can identify the high EQ intensity area and where the
source is, the removal of the source can be done. If the source cannot be removed, it
should be shielded with a heavy material or buried in the ground. Lead (Pb) is often
used as the shield. Concrete blocks (without holes are better) and sandbags are also
used. Hence, if some EQ sources are found on the ground surface, the effective way
to reduce the exposure is to remove, shield, or bury them.

If EQ sources are uniformly distributed or EQ intensity is uniformly high, the EQ
sources must be dispersed in air or water, or a very strong source exists at longer
distance (γ–photons can reach over 100 m in the air). Then, there is no other way
than escaping. Since the dispersed sources fall down with the rain, EQ intensity in
air usually becomes higher in a rainy day (time) than in a fine day (time) as seen
in Fig. 2.1 in Chap. 2. In any case, EQ intensity in air should be monitored, and if
the source was found, it should be removed. If the sources were deposited on the
surface of some structures, it should be removed with adhesive tape, etc. and/or wash
out with water. Of course, the adhesive tape and water used to remove the source
become contaminated and hence carefully stored until the radiation level decreases
below the regulation level or disposed of according to the regulation law.

5.3 External and Internal Exposures

Although internal and external exposures are briefly described in Sect. 2.5 in Chap. 2,
here is given a little more detail from the viewpoint of EQ exposure to a human body.

There is a fundamental question; whether the internal exposure is the same or
different from the internal exposure? In principle, there should be no difference in
the effects between the two for the same absorbed dose equivalent in Sv. However, it is
hardly possible to give the same absorbed dose equivalent to a tissue or an organ with
the both exposures. For the external exposure, some energy is lost in tissues passes
through before EQ arrives at a concerned organ. The loss is significantly different
among the kind of EQ. In case of the exposure to 131-Iodine (^{131}I) of which exposure
to thyroid is concerned, the internal and external exposures are significantly different.
^{131}I radiates both of 0.6 MeV of β–particles and 0.365 MeV of γ–photons. For
the external exposure, the β–particles do not arrive at the thyroid. Consequently,

the internal exposure to thyroid by ^{131}I is much more significant than the external exposure. In addition, because the β–particles give larger energy in small volumes compared to γ–photons, the absorbed dose given by the β–particles is far larger than that of the γ–photons. Hence, the internal exposure by RI emitting β–particles and/or α–particles is much dangerous than that given by RI emitting γ–photons.

Since a pocket dosimeter is calibrated to give absorbed dose equivalent or effective dose (in Sv) mainly for γ–photons converting the detected EQ intensity (cps or cpm) to the absorbed dose equivalent or the effective dose for a human body averaging the different effects given to individual organs (see Sect. 2.5.4 in Chap. 2), it is no use for the estimation of the internal exposure. It requires a full body counter to find EQ sources taken in a human body. In case of the exposure to ^{131}I and ^{137}Cs, which dominate radioactive fallout of the Fukushima accident and emit both β –particles and γ–photons simultaneously, the absorbed dose equivalent given by the γ–photons can be determined by a pocket dosimeter. Since the emitting ratio of the β–particles and γ–photons is fixed, the dose equivalent given by the β–particles can be estimated from that given by the γ–photons.

For the different kinds of EQ exposures with the same absorbed dose equivalent, the difference in appearing of their effects should be basically little. However, for the comparison between the external and internal exposures, the former would give less influence. In addition, for the internal exposure, it is difficult to remove the source so that the internal exposure tends to continue unless the source is removed.

In any way, to reduce the exposure, required is "to stay away from the source, or to keep the source away (shield)". For the reduction of the internal exposure given by the source taken into a body, the source should be discharged or exhausted with the assistance of appropriate chemicals or drugs. In case of tritium (T), diking water is effective. Of course, the best way is not to take the source into a body. In the next section, the reduction of the exposure is considered more detail.

5.4 Reduction of EQ Exposure to a Human Body

Small radioactive particles suspended in the air or adherent to foods are easily taken into a body by oral ingestion through nose and mouse. Although radioactive materials are also taken in a body through skins, their possibility is far less than those taken by the oral ingestion, except for the isotopic exchange of tritium (T) in tritiated water with H in water on the skin. In the case of the oral ingestion, the radioactive particles are delivered to and stored at tissues and organs through lungs or digestive systems, i.e., they are carried by blood or various body fluids to other organs in the body and stored. Depending on chemical forms of radioactive materials, they accumulated and are concentrated in a particular organ, like the accumulation of ^{131}I in thyroid, which is the main cause of the internal exposure. And the mechanisms of injection and ejection for Cs and I (the dominant radioactive fallout of the Fukushima accident) have been extensively studied. They are, for the injection, on how the radioactive materials are taken into the body organs, how transported via body fluids, and how

Table 5.1 Half-lives of disintegration and biological exhaust for dominant radioisotopes released at the Fukushima nuclear accident

Radioisotope	Half-life	
	Disintegration	Biological
^{131}I	8.02 days	120 days in thyroid
^{129}I	15.7 million years	12 days in other organs
^{137}Cs	30.1 years	~70 days
^{134}Cs	2.06 years	100–200 days
^{90}Sr	28.6 years	~49.3 years in bones
^{3}H	12.32 years	~10 days in tissues
^{32}P	14.3 days	

accumulated in organs, and for the ejection, how long is their residence time in the organs, and how they are excreted from the organs.

Regardless of the mechanisms, RI taken into an organ or tissue will be gradually exhausted when its non-radioactive isotopes are taken into the organ and replaced with radioactive ones by isotopic exchange. Some chemicals which have similar chemical nature as RI taken into the organ could also replace the latter. Usually, most of elements consisting a human body are replaced by the same chemical elements newly taken in. Accordingly, RI once taken into a body is gradually ejected with the biological half-life, i.e., the ejection rate is proportional to the amount remaining in the organ or the remaining amount decreases exponentially with time, and the time required to reduce in half is the biological half-life. The biological half-life varies from an element to element. Table 5.1 summarizes the biological half-lives for important RIs in a human body, together with their disintegration half-lives (decay time).

After the Chernobyl nuclear accident, patients of thyroid cancer have increased mainly in children. The thyroid is the organ that makes thyroid hormone and requires iodine (I) to make the hormone. Since I is not an abundant element in nature, I once taken into the body is easily captured in the thyroid. The Chernobyl accident dispersed ^{131}I in wide fields. It was taken into children's bodies from food, etc. and was concentrated in the thyroid to give internal exposure resulting cancerized thyroid tissue. There are two types of radioactive iodine, as shown in Table 4.1 in Chap. 4. Shorter half-life ^{131}I is much hazardous than long half-life ^{129}I. Because RIs with a short half-life give higher absorbed dose rate (i.e., higher emission rate of EQ). RIs with long half-life, on the other hand, continue to release energy giving lower absorbed dose rate over a long period of time. It is well known that administering iodine tablets before EQ exposure reduces the radiation damage caused by the radioactive I. The idea of the administering of the iodine tablets to protection of radiation hazards in the occasion of nuclear accidents is given in websites such as [1] and [2] as examples. The iodine tablet is consisting of Potassium Iodine (KI) containing only the stable isotope of ^{127}I. With its advanced administration, organs especially thyroid have prevented the take-up of the radioactive I. This is highly effective, particularly, in the occasion of nuclear accidents. Even after taking the radioactive I (^{131}I and ^{129}I) in, the continuous administration of the tablets accelerates isotope exchange to replace the radioactive I with a stable one and shorten its biological half-life. Drinking milk is also effective because it contains a lot of iodine. (Of course, the milk must be free from radioactive I). It is said that the side effects of I are not appreciable. However, there is a possibility of allergic reactions, and it is not recommended to drink them too much in a daily life. Nearly 10 years have passed since the Fukushima nuclear accident, and the concentration of ^{131}I has decayed to be 10^{-5}, so that the effects of ^{137}Cs become greater. Still, attention should be paid on selective injection of I in children.

Another important radioactive element is tritium (T or ^{3}H), which was not considered seriously just after the accident. T is easily taken into the body as an essential water and hence needs to be watched carefully. To remove T taken into the body, water intake is recommended. In particular, beers or some alcohols are effective. Water taken in promotes T discharge from the body.

In the Fukushima accident, seawater that was used to cool the reactors continues to be stored. Although most of radioactive isotopes contained in the stored water were removed to be far below the regulation levels, T is so hard to fully remove because it is converted to HTO with its easy isotopic replacement with H in H_2O and the concentration of HTO was very low. In principle, removal of T from water is possible, but its cost is enormous. Since the amount of T contained in the stored water is so tinny that no influence is expected if it is released into open sea. Nevertheless, due to concerns on damages caused by harmful rumors, it is not decided to release it at present. Scary feeling to radiation without its understanding likely forces to hesitate the release. Considering that the cost and risks are always accompanied in use of any engineering or technology, the judgment to allow or not allow the release should be hopefully done on scientific base.

T and ^{32}P appeared in lower columns of Table 5.1 and are often used in medical purposes. Owing to their activity, their profiles in the organ can be determined which make it possible to distinguish abnormal site like cancerized tissues from normal tissues. Although their exposure effects are concerned, the internal exposure could be kept low owing to the short decay time of ^{32}P or the shorter biological half-life of T.

Again, the only way to prevent the internal exposure is to avoid the intake of EQ sources. Once the source is taken in, it should be removed as early as possible, with enhancement of its ejection. To do this one should know where and what kind of RIs are accumulated in an organ or body. In this respect, detection and measurement of EQ are dispensable. When the location and kinds of RIs are identified, focused treatment to remove them becomes possible with such like the shortening of the biological half-life. As described above, the basic way is to replace an RI with a stable isotope of the same element (iodine is an example). Alternatively, some chemical agents which have similar chemical properties with the subject including RI but are harmless to the human body can be used to replace the latter. For example, since Cs is an alkaline metal, its family members in the chemical table, lithium (Li), sodium (Na), potassium (K), rubidium (Rb), and francium (Fr) can replace ^{137}Cs. There is another method using chemical reactions by some chemical compounds that are easy to bind to a specified RI.

^{90}Sr released with the Fukushima accident and remained long is concerned to cause bone cancer because it is likely accumulated in bones like Ca which as an alkaline earth element behaves similarly with Sr. In this case, Ca intake should be useful, but it is still under research on how it can be taken into the bones.

For the removal of EQ sources taken into a human body, various reports and websites are available, for example see a review "Decontamination of RIs" [3, 4].

5.5 Resilience

5.5.1 Where and How Large Areas Are Damaged or Influence by EQ Exposure?

In Chap. 3 are described material damages and recovery caused by EQ exposure. Here are described again focusing damaging and recovery of a human body exposed to EQ. EQ injected into the human body lose their energy through collisions with electrons and nuclei. There is no specialty. The collisions with nuclei displace atoms from their fixed positions in various organic molecules including DNA resulting in direct break of chemical bonds or introduction of defects in the molecules, and those with electrons also cause similar events but indirectly through electron excitation. Defects and damages given by the EQ exposure in the human body are mostly caused by the electron excitation, with small contribution to the nuclear collisions. Although most of the defects or damages once generated are recovered, small portion of defects and damages initially generated remains long after the EQ exposure. The volume of damaged or influenced area by the EQ exposure is quite small around a few nm in radius along the trajectories of EQ, as shown in Fig. 4.13 in Chap. 4. Therefore, the influence of EQ exposure or appearance of the damaged area varies significantly depending on where EQ hit or passed through in cells, tissues, organs in a body, and on the kinds of EQ.

As discussed in Chap. 4, the G value, defined as the number of broken water molecules with the EQ exposure giving 100 eV into water is about 8. Then it can be estimated how many H_2O molecules are decomposed in a human body (60 kg) exposed to 1 Gy (1 Sv) which gives some influences as follows. The total exposed energy is 60 J. Suppose 70% of a human body is consisting water, the total absorbed energy in water part is 42 J, which is equivalent to $42 J \times (1.6 \times 10^{19}) eV/J = 6.7 \times 10^{20}$ eV. Then the number of decomposed water molecules is $6.7 \times 10^{20} \times 8 \div 100 = 5.4 \times 10^{17}$ molecules $= 1.3 \times 10^{-6}$ mol or 2.4×10^{-5} g (24 mg). Compared with the whole weight of 60 kg, 24 mg is only 0.4 ppm, which is a surprisingly small number.

Since the atoms and radicals produced by the water radiolysis are chemically active, they could cause abnormal chemicals which do not normally appear in cells and tissues. If the decomposition of water occurs near important molecular species related to the proliferation or metabolism of biological cells such as DNA or RNA, their influence would appear as the death or canceration of the cells. While the influence of the decomposition appeared far from the important species would be quite small. One cannot determine where the unwanted chemical reaction or decomposition appears, which is quite statistical. i.e., the appearance of the EQ exposure effects is essentially probabilistic for the EQ exposure with lower absorbed dose.

Furthermore, it should be noted that because the cells and tissues have resilience from the damages, the remaining effects become basically milder. In addition, the importance of tissues in a body varies greatly from tissue to tissue, irradiation effects are greatly different depending on the location of energy deposited area in nanometer scale, as discussed in Sect. 4.4 in Chap. 4 and on what kind of cells or tissues of organs are there.

Again, the effects of lower absorbed dose exposure on the human body (appearance of diseases) appear inherently stochastics or probabilistic. In an extreme case, exposure to just one energetic quantum could destroy DNA in a quite important cell, while a higher absorbed dose exposure to the area far from its DNA in a cell would not give significant effects. Usually, around 100 mSv is critical dose for appearance of the effect of EQ exposure. However, the critical dose largely deviates depending on individuals, or their living conditions. Someone would be influenced by such small dose of 1 mSv, while some could be tolerant over 1000 mSv (1 Sv).

Thus, in the case of the external exposure, the influence varies greatly depending on the location of EQ injected, while for the internal exposure it depends on the location of the EQ source, i.e., the damage is concentrated in an organ or a tissue retaining the EQ source. Still, the influence or damage caused by the EQ exposure appears differently depending on resilience of individuals, which is discussed in the next section.

5.5.2 Recovery of Damages and Resilience

In the previous section, most of initially produced defects/damages by EQ exposure in inorganic materials are immediately or in very short times recovered and only their small portion remained as radiation damages or defects. In addition, the living beings have functions to recover the damages referred to as resilience as described above. In Fig. 5.2 [5], damaging and recovery of DNA are schematically illustrated.

The resilience has not been quantitatively evaluated. Basically, it is similar to the resistance to disease which varies from person to person. It is well known that mental strength or health is a quite important factor for the resilience to flu for example. Nevertheless, it is not clarified where the mental strength or resilience originates in brain. The resilience varies greatly from person to person, the mental state, and surrounding environment of the individuals. If one can stay in good health and mental conditions or in comfortable surroundings, his possibility to have stronger resilience or power of recovery from the radiation effects would be high.

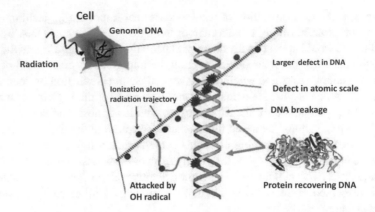

Fig. 5.2 Models for damaging and recovery of DNA [in [5], Reprinted with permission]

It is also important to note that the resilience appears in various stages from damages in atomic scale, DNA, cells, tissues, organs, and a whole body. Usually, when DNA in a cell is damaged, the cell loses its regeneration function resulting in its death. However, tissues have the function of eliminating dead cells, the death of individual cells in them does not immediately lead to the death or canceration of the tissue and organ including the dead cell.

If one is happened to be exposed to EQ by a nuclear accident, he cannot reduce the absorbed dose. However, the damage caused by the EQ exposure could be mitigated if he believed in resilience. Living with resilience but without depressed mind will reduce the effects of the exposure.

As mentioned in Chap. 1, one air round trip from Tokyo to New York will give the absorbed dose equivalent of approximately 0.2 mSv. Although businessmen who travel back and forth 10 times a year are exposed to 2 mSv per year, they are rarely concerned about the influence of the exposure on their health. The effects of jet lag and business stress would give more influence. Hence even if someone got some health damage, it would not easy to find the cause either of which the exposure, the stress or some others. Instead, active involvement in their jobs could give more resilience owing to their mental power compared to normal people.

Unfortunately, after EQ exposure, one cannot reduce the absorbed dose. The mental depressing owing to the exposure could reduce the resilience to exposure damages. Absorbed dose equivalent for most of people in nuclear accidents including the Fukushima is not high enough to give immediate irradiation effects. Of course, it cannot be excluded that some impact would appear in the future, but without additional exposure, the impact of the EQ exposure so far would remain quite low in the future. Instead, comparable or more health risks would be given by pesticides, tobacco, food, and life stress for most people. It is also true that similar effects caused by DNA damage on EQ exposure could be given by medical or drugs like thalidomide, DDTs, dioxins, substances called environmental hormones, etc. through some chemical reactions in DNA. Active and comfortable daily life believing resilience would be the best way to live long.

5.6 Short-Term and Long-Term Exposure

It should be noted that both the values of absorbed dose, Gy, or absorbed dose equivalent, Sv, are integrated values for certain time duration of EQ exposure, usually a day, a month, or a year. Since EQ intensity in nature is quite low, they are expressed in the unit of μGy or μSv. In actual EQ exposure, the effects appearing after the exposure of the same absorbed dose equivalent but different time duration of 1 h or 1 year would be different. However, it is not always possible to tell what differences will appear. Although there are comparative studies for the exposure of high absorbed dose rates where the effects are revealed in animal studies, for lower absorbed dose rates it is difficult to find accurate data unless a huge number of individuals are examined.

As described in the previous section, in any substance, there appears recovery of the initial damages given by EQ exposure. Those damages escaping from the recovery could interact with each other and turned into be stable damages, like the changes of interstitials to their clusters and loops in inorganic materials. Owing to different recoveries, the final damages are significantly different with characters of exposed substances. In addition, both the initial damages and recovery are different depending on absorbed dose rates. For lower dose rates, initial damages should appear discretely or independently from each other, and the recovery should be also independent. Accordingly, the irradiation effects would be independent of the dose rate. Under higher absorbed dose rate exposure, possible overlap of the initial damages would induce their accumulation resulting in different kinds of damages from those appeared for the lower dose rate so that different recovery processes would appear. Therefore, under the same absorbed doses, the effect would be higher for the exposure with higher dose rate (exposure within shorter time) than that with the lower dose rate (exposure with longer time). Furthermore, the effects for longer time exposure with lower dose rate are quite hard to predict particularly for living beings because of their resilience in long time owing to their metabolism.

From these points of view, it would be better to discuss the irradiation effects based on the deposited power (G s^{-1} or W kg^{-1}) and its integration of total absorbed energy (J kg^{-1}) instead of the absorbed dose equivalent or effective dose (Sv) usually used. Still, the information on how long it has been exposed, what kind of EQ is exposed, and how high the energy of EQ is, are missing in the absorbed energy or dose.

Historically, the understanding of radiation has changed, from the radiation as something unknown, as carrying energy, and as consisting of different kinds of EQ. Their hazardousness was recognized as diseases appeared in living beings after the EQ exposure. In early days of radiation biology, the effects of the irradiation appeared, such as a sense of fatigue, burn, dehairing, cancer, and so on, were summarized with

Table 5.2 Relations of units used in discussions on EQ exposure

SI unit	Conventional (CGS) unit	
Intensity of radiation source (Radioactivity)		
Bq (Becquerel)	**Ci** (Currie)	
t^{-1}	$1\ Ci = 3.7 \times 10^{10}\ Bq$	
Absorbed energy		
J/kg		
Absorbed dose	Absorbed dose	**R**adiation **a**bsorbed **d**ose
Gy (Gray)	**R** (Roentogen)	**rad**
$1\ Gy = 1\ J/kg$	$1\ R = 8.77 \times 10^{-3}\ Gy$	$1\ rad = 10^{-2}\ Gy$
Absorbed dose equivalent	Roentgen equivalent **man**	
Sv (Sievert)	**Rem**	
$1\ Sv = W_R \times 1\ Gy$	$1\ rem = 10^{-2}\ Sv$	

Wr: Radiation weighting factor (see Table 2.1 in Chap. 2)

using conventional (CGS) units different from the present SI unit, i.e., for absorbed dose, Roentgen (**R**) (1 R = 8.77 mGy), Radiation Absorbed Dose (**rad**) (1 rad = 10^{-2} Gy), and for absorbed dose equivalent, Roentgen Equivalent Man (**rem**) (1 rem = 10^{-2} Sv). To distinguish the different effects causes by different kinds of EQ and differences of exposed substances (skins, tissues, organs), the effective dose which is also represented in Sv was introduced. Thus, the conventional units of **R**, **rad,** and **rem** are now redefined as Gy and Sv in SI units. Relations of these units are summarized in Table 5.2.

Anyhow, most of EQ exposures that induce the radiation effects on living beings are due to β–electrons or γ–photons. Consequently, there is no large difference in absorbed dose (Gy) and dose equivalent (Sv). One can claim that it is impossible to summarize complex radiation effects with only the amount of the absorbed dose or dose equivalent. Now, owing to recent advances in understanding of EQ (radiation), it is possible to distinguish the kind of EQ and to measure their intensity and energy. This book entitled "Radiation: An Energy Carrier" could not be published several decades ago.

In order to avoid radiation hazards, regulation levels of EQ exposure are separately defined for ordinary people and professional workers. The allowable absorbed dose equivalent for the ordinary people is usually 1 mSv for one year, and in occasion of the emergency, it is set to be within 20 to 100 mSv for one year. For the professionals, the regulation levels are 1 Sv, 100 mSv, and 50 mSv for a whole life, 5 years, and a year, respectively. Somewhat lower levels for the effective dose are set for radiation-sensitive organs, like eyes, skins, and so on. More detailed regulations in an hour or a day are introduced for workplaces possibly giving higher absorbed dose equivalent.

Although the regulation levels are for integrated dose within the fixed day/time assuming the constant absorbed dose rate for a day or a week, the limit for instantaneous exposure or the maximum allowable dose to be received at one time is not specified. The occasion of spontaneous exposure with quite high absorbed dose must be limited in accident. Spontaneous exposure over 1 Sv would give clear damages as indicated in Table 2.3 in Chap. 2.

If a worker were exposed over the regulation level before allowed time to work, an hour, a day, a week, etc. he should be off from his work for the remaining time. In case that one is asked to work at high dose rate area, his working time should be limited beforehand not to be exposed over the regulation level. In Fukushima and other areas subjected to nuclear accidents, workers are handling radioactive materials in rotation.

References

1. http://www.jodblockade.de/en/iodine-blockade/
2. http://www.webmd.com/vitamins/ai/ingredientmono-35/iodine
3. L. Domínguez-Gadea, L. Cerezo, Decontamination of Radioisotopes, Reports of practical oncology and radiotherapy 16(4), 147–152 (2011)
4. https://www.cdc.gov/nceh/radiation/decontamination.html
5. http://www.asrc.jaea.go.jp/soshiki/gr/eng/mysite6/index.html

Chapter 6
Detection and Measurement of EQ (Radiation)

Abstract In order to know the existence of EQ (radiation) and to assess the effects of their exposure, the type of EQ (α-, β- and β^+-particles, γ-photon, X-ray, neutron, and other quanta) should be identified and their intensity (Bq) and energy should be measured. In this chapter, basic physics of EQ detection and methods for measurements of intensity and energy of EQ are described. Once EQ are identified and their energy and intensity are measured, absorbed dose (Gy) and absorbed dose equivalent (Sv) in a substance exposed to EQ can be estimated.

Keywords Absorbed dose · Detection · Dose equivalant · EQ intensit · EQ energy · Gray Measurement Sievert

6.1 Introduction

In order to know the existence of EQ (radiation) and to assess the effects of their exposure, the type of EQ (α-, β^-- and β^+-particles, γ-photon, X-ray, neutron, and other quanta) should be identified and their intensity (Bq) and energy should be determined. Once they are identified and determined, absorbed dose (Gy) and absorbed dose equivalent (Sv) in a substance exposed to EQ can be estimated. In this book, it is often claimed that the detection of EQ is easy. However, it is not easy to distinguish the kind of EQ and to determine their intensity and energy.

There are two kinds of radiation detectors or radiometers. One can measure only the intensity of EQ, while the other can measure both intensity and energy, simultaneously. Table 6.1 summarizes principles of detection, names of detectors and the kind of EQ to be measured. What the detector can determine or measure depends on its physical bases. This section, focuses on the measurements of β-particle and γ-photons, which have a larger impact on a human body with the external exposure.

These detectors provide information about the intensity (number of EQ in cps or Bq) and the energy of EQ. However, they do not give how much absorbed dose or dose equivalent a person has got by the EQ exposure, and the results of measurements should be converted to the absorbed dose or dose equivalent as described in Chap. 2. Those detectors calibrated to give the absorbed dose and/or dose equivalent for a person are referred as dosimeters. Following are brief descriptions of measurements

© Kyushu University Press 2022
T. Tanabe, *Radiation: An Energy Carrier*,
https://doi.org/10.1007/978-981-19-1957-2_6

Table 6.1 Radiation detectors

Principle of detection		Name of detector	EQ to be measured
Ionization	Gas	Ionization chamber	α-particle, β-particle, γ-photon
		GM counter	β-particle, γ-photon
		Proportional counter	β-particle, γ-photon, neutron
		Gas-flow counter	β-particle, γ-photon
	Solid	Semiconductor detector	β-particle, γ-photon, X-ray
Scintillation		NaI (Tl) scintillation detector	γ-photon
		ZnS (Ag) scintillation detector	α-particle
		Plastic scintillation detector	β-particle
		Thermoluminescent dosimeter (TLD)	γ-photon, X-ray
		Photon glass dosimeter Glass badge)	β-particle, γ-photon, X-ray, neutron
Photograph		Film badge	β-particle, γ-photon, X-ray, neutron
		X-ray film, imaging plate	α-particle, β-particle, γ-photon, X-ray, neutron
Energy measurement		Calorimeter	α-particle, β-particle, γ-photon

of the intensity and energy of EQ in Sect. 6.2 and on determination of the absorbed dose and dose equivalent in Sect. 6.3. In Sect. 6.4, visualization of the distribution of EQ sources is introduced, which has recently progressed significantly.

6.2   Determination of Type, Intensity, and Energy of EQ

EQ detectors listed in Table 6.1 are separated into two groups; one can measure only the intensity of EQ injected into the detector, and the other can distinguish kinds of EQ and determine energy deposited or absorbed in the detector. Since various textbooks describing the basic physics for radiation measurement are available, here are briefly introduced EQ measurements.

6.2.1   Measurements of EQ Intensity

The simplest detector of radiation or EQ is a GM (Geiger-Müller) counter or tube. Figure 6.1 is a photograph of typical one, a battery-powered GM counter. It is based on the ionization effects of EQ as described in Fig. 6.2. In a GM counter, there are two electrodes, one, anode at the center, and the other, cathode surrounding the anode with inert gas like He, Ne, or Ar enclosed. When one energetic quantum enters in the counter, the inert gas molecules are ionized to create positively charged ions and electrons. A high voltage is applied between the anode and the cathode to attract and accelerate ions and electrons to the cathode and the anode, respectively. During the movement of the ions toward the cathode and the electrons toward the cathode, they obtain acceleration energy and repeatedly collide with inert gas atoms to give

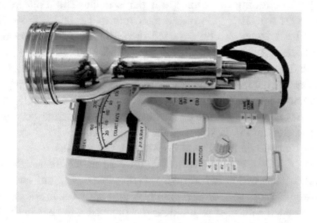

Fig. 6.1 Photograph of a battery-powered GM counter

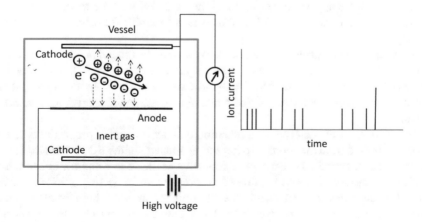

Fig. 6.2 Principle of operation of a GM counter and output current pulses

energy and ionize them cascadingly. Thus, large number of ions and electrons are formed in the counter. First electrons reach the anode, and a little later, ions reach the cathode. All of these charges are measured as an electric current with a pulsed shape as shown in the right figure of Fig. 6.2. The number of current pulses thus measured corresponds to the number of EQ entering the counter as counts per unit time, i.e., cps, cpm, or cph (counts per second, minute, or hour, respectively), and the height of the pulsed current in Fig. 6.2 correlates with the energy of each quantum entered in the counter.

Electric current (i) is defined by charges (Q) that move per second.

$$i = -dQ/dt. \tag{6.1}$$

The current of 1 ampere (A) is defined as the current generated by the charge of 1 coulomb (C) entering or coming out per second. Thus, one electron or ion having charge of 1.6×10^{-19} C gives the current of 1.6×10^{-19} A. Such a small current cannot be detected. To be detected the current should be over 10^{-10} A. In a GM counter, electric charges are multiplied more than 10^9 times. Although the atmosphere are full of electromagnetic waves (such as visible light and radio waves) which are basically quanta, their energy is too small to ionize the inert gas in the GM counter. It should be noted that the GM counter can detect basically one energetic quantum whichever it is a photon or particle, while the actual current given by a single energetic quantum is as small as the order of 10^{-18}–10^{-19} A. This is one of the reasons for that people fear EQ, not visible but hazardous.

Since most of EQ in the atmosphere is emitted at the disintegration or decay of radioisotopes (RIs) which is a random phenomenon, EQ emission or detection sequence in time is random. Correspondingly, the random current pulses appear in the GM counter. As the number of incident EQ increases, the gas in the GM counter is fully ionized which makes the pulse height saturated and disables subsequent measurement resulting in dead time of detection. Therefore, when the EQ intensity is very high, there appears count-loss and the pulse height is no more proportional to EQ energy. Accordingly, the GM counter is not suitable for determination of the EQ energy.

In a GM counter, current pulses are often converted to sound pulses so that every sound pulse corresponds to EQ entering the counter. For intense EQ, the interval of sound pulses becomes shorter and the sound changes to quacking. So, some people may misunderstand that the simplified name of a Geiger counter originates from the sound.

The number of EQ detected in a radiation counter is roughly proportional to EQ intensity emitted from the source. To get actual intensity of the EQ source (Bq kg^{-1}), calibration is required considering the geometry of the source and the counter, i.e., solid angle, surface area, and volume of the counter, and detecting efficiency of the counter as described in Sect. 6.2.4. A GM counter can detect both β-particles and γ-photons, and is more sensitive to the latter than the former because β-particles which do not have enough energy to pass through the window of the GM counter cannot be detected. In addition, the GM counter cannot specify the kind of EQ, either β-particles or γ-photons, nor determine EQ energy, because the pulse height is easily saturated losing correlation with the energy of EQ.

Recent advances in electronic technology, especially for semiconductors, have led to the introduction of semiconductor detectors that use EQ to excite electrons into their conduction band instead of ionization of gases in a GM counter. As described in the next section, the use of Germanium (Ge) as a detector of the semiconductor detector makes the current pulses of which height is proportional to the energy of EQ.

6.2.2 Accuracy in Intensity Measurements

The intensity of EQ emitted from RI (radioactivity of RI) exponentially decreases with its half-life, while the disintegration of RI occurs randomly. Therefore, in EQ detection, the time interval between succeeding two pulses changes randomly as shown in Fig. 6.2. A measurement of natural radiation gives around 10 cps or 100 cpm. In repetition of the measurements, the counted number randomly changes a little on each measurement in cpm, like 98, 110, 80, 101, and so on. If the averaged count was 100 cpm after multiple measurements, scattering of the measured counts would be mostly within 10%, i.e., the count number would be between 90 and 110. However, it could be occasionally 120. If you increased the counting time, for example, 30 min, the average counts would be around 3000 with scattering of some 5%. In both cases, the counts in a minute are the same as 100 cpm. However, the scattering or deviation from the average becomes smaller for longer time measurements. In other words, the accuracy of the intensity measurements of EQ becomes better for longer time measurements.

The nuclear disintegration is a statistical phenomenon, and the deviation of the number of the disintegration in a unit time follows the Poisson distribution in the case of lower number of the disintegration, while the Gaussian distribution for the higher number. Following the Poisson distribution with the averaged number of decays, M, the deviation from the average is given $\sqrt{M}$. Therefore, in a case with detected number of N, the deviation from N would be $\sqrt{N}$ and represented as

$$N = N \pm \sqrt{N} \tag{6.2}$$

In case of following the Gaussian distribution, the deviation in repetitive measurements would be two or three times the standard deviation (σ) from averaged value of N as given by,

$$N = N \pm 3\sqrt{\sigma} \tag{6.3}$$

As such, radiation measurements always include some error or deviation. The causes are mostly the statistical nature of nuclear disintegration and some errors in electrical systems of a detector. The latter could be technically as small as possible.

Although the statistical deviation can be reduced with increasing the measuring time, the scattering is inevitable and there is no way to remove it fully. It should be reminded that in measurements of lower intensity EQ of near-natural EQ level, around 100 cpm, the measured value always includes error of around 10%.

6.2.3 Measurements of EQ Energy

EQ (Radiation) detectors called a scintillation counter and a semiconductor detector can determine the energy of EQ. They measure deposited energy of EQ together with the number of EQ entered in the detector. Hence the data obtained by them are expressed as an energy spectrum or energy profile of EQ with energy as the horizontal axis and the number of EQ detected as the vertical axis as shown in Fig. 6.3 [1], which is the energy spectra of radioactive materials in fallout of the Fukushima nuclear reactor accident sampled at 4 days and 8 days after the accident together with background. The energy spectra are consisting of continuously decreasing ones in energy ranging from 0 to over 2.5 MeV and sharp peaks (referred to as line spectra) originated from the disintegration of RIs indicated in the figure. Considering that the vertical scale is given in logarithm, the intensities of the line spectra are higher than the continuous spectra by 1–3 orders of magnitude. All RIs except ^{40}K in the figure are fission products (FPs) of nuclear reactions. The reduction of the continuous

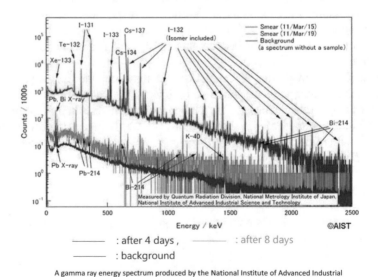

Fig. 6.3 Energy distribution of EQ emitted from fallouts released at the accident of Fukushima nuclear power plants. The fallouts were collected by smearing at 4 days and 8 days after the accident. The blue line is the measurement result of the background (no sample). ^{40}K is originated in nature. Reprinted from [1] licensed under "Copyright Miraikan" "© Miraikan"

spectra is significant for only 4 days (from May 15 to 19), while the reduction of the line spectra corresponds to the half-life of respective RI. ^{132}Te which has the shortest lifetime among all FPs detected was appreciable only in earlier times. The line spectra given by ^{132}I also decreased rapidly. Intensive line spectra given by ^{131}I, ^{134}Cs, and ^{137}Cs have remained long. Among four peaks in the spectra given by ^{131}I, their relative peak intensity ratios are just the same as those given by the disintegration of isolated ^{131}I, as 0.284 MeV (6.14%), 0.365 MeV (81.7%), 0.637 MeV (7.17%) given in Fig. 3.4 in Chap. 3. After decaying the dominant lines, minor ones like ^{214}Pb and ^{214}Bi became clear. The peak at 0.662 MeV is due to the γ-decay of ^{137}Cs, which now remains as the most dominant after 10 years of the accident.

6.2.4 Calorimetry

The energy carried by EQ is finally converted into heat in the substance exposed to EQ. A calorimeter is an EQ detector using this physics in which all energy carried by EQ is converted to heat and is detected as a temperature rise of a thermally isolated substance exposed to EQ. Specifically, an EQ source is set at the center fully surrounded by a heat sink material with thickness enough to stop all EQ in it, and thermally isolated. Then the temperature rise of the heat sink material gives the deposited (absorbed) energy (E) as

$$E = m \times C \times \Delta T / \Delta t, \tag{6.4}$$

where m and C are respectively the mass and the heat capacity of the heat sink material and ΔT, the temperature rise during the time duration of measurement, Δt. The calorimeter is used as one of the most important methods to determine absolute decay heat of or released energy from RI.

6.2.5 Intensity (Radioactivity) of EQ Source

EQ emitted from point or planer sources is not homogeneously distributed in space. The vertical axis in Fig. 6.3 shows counted numbers of EQ by a semiconductor detector over 1000 s. Although the counted numbers are proportional to the emitted numbers of EQ from the source, they do not indicate the intensity (radioactivity) of the EQ sources in unit time and unit mass, i.e., Bq kg^{-1}. To get the EQ intensity of the source, the number of counts measured must be converted considering the geometrical relationships of the source and detector or a solid angle of the source to the detector and the detecting efficiency of the detector. Figure 6.4 [2] shows the tracks of α particles from thorium ($C + C'$) in a Wilson chamber, showing the two ranges. The α particles collide with molecules in the air and lose energy, so they can penetrate in very short distance (less than 1 mm) even in the air as indicated in the tracks. In contrast, γ-photons travel or penetrate much longer distances, more

Fig. 6.4 Tracks of α
particles from thorium ($C +$
C') in a Wilson chamber,
showing the two ranges. In
[2], reprinted with
permission from Cambridge
University Press

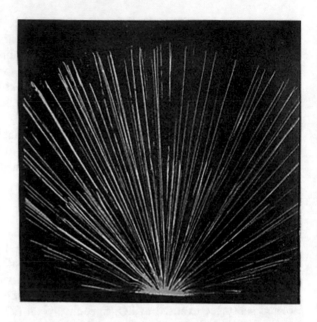

than 10 m in the air, for example. Since the sensing area or volume of the detector
is limited, the counted (detected) number of EQ decreases with the inverse square
of the distance from the source to the detector for a point source or the very small
volume source, while the inverse of the distance for a planer source.

Figure 6.5 shows how the intensity of EQ changes with the distance from the
source to the detector. In the figure, $r \cdot d\theta$ is a sloid angle of which surface area

Fig. 6.5 Schematic of EQ
emission from a point source
at the center of a sphere. The
number of EQ passing per
unit area proportionally
increases with its solid angle,
$r \cdot \sin\theta \cdot d\varphi$, and decreases
with inverse of square of
radius, r^2

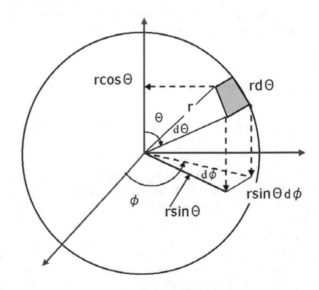

increases with the square of the distance from the source to the surface. Suppose no attenuation of EQ in air, the number of emitted EQ passing through a fixed solid angle is constant. Hence the detected number of EQ from the point source by a detector with a certain detection area decreases with the inverse of the solid angle or square of the distance.

In a detector, all EQ entered are not necessarily detected, or there always appear counting losses. Therefore, to get real number of EQ entered in the detector should be calibrated with its detection efficiency (measured counts over number of EQ entered). To determine the flux (particles $m^{-2} t^{-1}$) of EQ passing through the location of the detector, the measured value should be corrected with the detection efficiency and the sensitive area of the detector. In case of a point source, its intensity of EQ (Bq) can be determined from the flux divided with the solid angle of the detector, or the surface area of the detector divided by a sphere with the radius equivalent to the distance from the source to the detector.

It was reported that some meet produced in Fukushima area included ^{137}Cs exceeding the regulation limit (500 Bq kg^{-1}). However, β-particles from the meat with the radioactivity of 500 Bq kg^{-1} give only around 2 cps when measured by a small GM counter having detection area of 10 cm^2 and the detection efficiency of 50% set at 10 cm from the center of the meet, as

$$0.5(\text{detection efficiency}) \times 500(\text{Bq}) \times 10((\text{detection area}) \div (4\pi \times 10^2)) = 2 \text{ cps}.$$

Therefore, even if injection of the meat containing Cs with the intensity equivalent to the regulation level would not give an immediate impact, or the injection of 1 kg of meet emitting 10 cps of EQ would not be hazardous.

Since EQ from the soil is considered to be released from planar sources, the dose in the upper space (air) of the soil decreases with inverse proportion to the distance (although it does not decrease much when the source area is enormously wide. Rather, it may increase a little above the soil.). Consider the measurements with a GM counter at little above the soil. The regulation level of Cs in soil is set at 5000 Bq kg^{-1}. Because the density of the soil is about 2, the soil of 1 kg has a volume of 500 cm^3 (a cube with 8 cm side). Suppose the regulation level of Cs is concentrated on the one surface. EQ are emitted to both upper and down directions from the surface area of 64 cm^3, resulting in

$$5000(\text{Bq}) \div 2(\text{direction}) \div 64 \text{ cm}^2(\text{surface area}) = 39 \text{ Bq cm}^{-2}$$

per unit area. Using the above-mentioned GM counter with the detection efficiency of 50% and sensitive area of about 10 cm^2 in close contact with the soil, the counting rate of $39 \times 10 \times 0.5 = 200$ cps will be detected. In reality, Cs are distributed inside of the soil, the counting rate would be less. In any way, if the counting rate just above the soil was over 100 cps, the EQ intensity (radioactivity) of the source in the soil would be above the regulation level.

Although above-mentioned two cases are just for rough estimation, if you would detect over 100 cps with a GM counter commercially available, you may need to ask your regulation authorities or local government to make a detailed measurement.

6.3   Absorbed Dose Measurements

As described in the previous section, simple EQ detectors like a GM counter give rough number of the intensity (radioactivity) of the EQ source after making appropriate calibration. However, simple detectors do not give energy of EQ, and hard to distinguish the kinds of EQ, which are necessary to determine absorbed dose and dose equivalent.

At the Fukushima accident, various RIs originated from nuclear fission products, and were dispersed in the atmosphere and on the ground surface. Accordingly, absorbed dose and dose equivalent for people in some areas near the Fukushima power plants exceeded far above the regulation level, and they were evacuated. For the safety, Japanese government decided allowable levels for living in a village or a town to be 20 mSv y^{-1} and 1 mSv y^{-1} for adults and children, respectively, which are a little less than the regulation levels decided in Japanese low. Although the relationship among the intensity and the energy of EQ and the absorbed dose equivalent is already described in Chap. 2, here is discussed again from the viewpoint of measurement.

Figure 6.6 is a photograph of a pocket dosimeter, which indicates absorbed dose equivalent (in μSv) of a person who wears it. By the way, as detailed in Chap. 1, "exposure" means that a substance (irrespective of living beings or inorganic matters) is exposed to EQ. Sometimes it is used limitedly to represent the energy deposition to living beings or particularly human-beings given by the exposure of EQ, as absorbed dose, Gy or J kg^{-1}, and absorbed dose equivalent, Sv.

In a dosimeter, as shown in Fig. 6.6, is set a program or conversion formula that changes absorbed (or deposited) energy measured in it to absorbed dose or dose equivalent in a human body. Therefore, when the intensity of EQ is uniform in a space where a person stays with the dosimeter, it gives a reliable value of the absorbed dose, while the intensity is nonuniform, such as given by a point source,

Fig. 6.6 Photograph of a pocket dosimeter

Fig. 6.7 A whole-body counter for determination of profiles of EQ in a human body

the dosimeter could give different values depending on the distance from the source and where the dosimeter was worn.

The dosimeter does not work for the internal exposure at all. For the internal exposure, it is necessary to find the location of a radiation source, kind of emitting EQ, their energy, and intensity. The only way to determine the location of the injected EQ source and the intensity and energy of emitted EQ in a human body is to use a whole-body counter shown in Fig. 6.7. Just like a PET scanner, counters are placed around the whole body and scanned from head to foot. However, if the α- or β-particle sources are injected into the body, it is quite hard to detect them because they are not able to come out from the body surface. RIs of Cs and I are dominant sources on a nuclear accident that emit both β-particles and γ-photons. The detection of the β-particles is hard, while γ-photons easier. Both RIs emit β-particles and the γ-photons with fixed ratio, the detection of the γ-photons which is easier allows to estimate the emission of the β-particles.

The source in the body emitting only α-particles like thorium given in Fig. 6.5 cannot be detected from the outside. However, the existence of RI in the body can be detected in-body fluids, i.e., blood, lymph, urine, etc. The exhaust or ejection of the in-body source could be monitored by the decrease of detected counts of them.

In any way, it is not easy to evaluate the absorbed dose and dose equivalent for a human body or a specified organ. Since there is no direct method of determining the absorbed energy or dose given by the internal exposure, there is no other way but to estimate based on where and what kind of the source is and how high its intensity is. The internal exposure could never happen if the source was not injected into the body. Hence one should be very careful with oral injection of any EQ sources. However, once it was taken into a body, it should be exhausted with the biological half-life. It is important to refrain from being too much sensitive or overreaction than necessary.

Focusing on the effects of EQ exposure or in the development of radiation biology, an absorbed dose equivalent or effective dose in the unit of Sievert (Sv) was introduced, so that it is convenient to evaluate and compare the effects of EQ exposure with them, while it is not easy to remind how EQ gives energy. In order to properly evaluate the effect of EQ exposure, it would be better to consider directly the type of EQ, its intensity, and energy, which are the base to define the absorbed dose and dose equivalent. Also important is which tissues or organs are exposed or concerned. Thus, the better way is instead of using Sv, getting correct absorbed energy given in Gy and predict the effects considering the kind of EQ and the exposed tissue or organ. In general, the values of Gy and Sv are not very much different from each other (for the irradiation of γ-photons and β-particles both are the same), and either can be used for a rough discussion. For precise discussion, Gy would be better. Unfortunately, however, data showing radiation effects have been historically arranged with Sv.

Anyhow, regulation levels of emergency for children is set at 20 mSv for a year, on which there may be argument either too low or too high. It should be noted that the value of 20 mSv is not very strict, but large uncertainty caused by statistical nature of EQ emission and resilience of organs do not distinguish the radiation effects caused by absorbed dose equivalent of 10–40 mSv, and difference among individuals would give larger difference, as described in Sect. 2.5 in Chap. 2.

6.4   Visualization of EQ Source Distribution

Except for extremely high absorbed doses of EQ exposure like over 1 Gy, the effects of EQ or EQ themselves are not visible nor sensible. Although radiometers and radiation counters detect EQ, they cannot indicate how EQ is distributed or radiated from the source. Recent advancements in EQ detection techniques and digital visualization techniques make the visualization of the distribution of the EQ source possible. The visualization technique is basically the same as used in X-ray inspection for medical purposes and non-destructive tests of structure materials and is referred to as X-ray fluoroscopy or X-ray transmission imaging. Energy deposited or absorbed in a substance or a human body exposed to X-rays differs depending on the density of the substance. Accordingly, the intensity distribution of transmitted X-rays reflects different densities of tissues or organs. The detection of the transmitted X-ray uses a photographic technique. In recent days, the conventional photographic technique has been replaced by a digital photographic technique as described in the next paragraph. Probably, most of the readers have been inspected with chest X-rays. In Japan, average absorbed dose given by the X-ray exposure for the medical purposes for individuals is 3.7 mSv per year, a little higher than world average.

EQ sources or materials emitting EQ are visualized as their intensity profiles, like an X-ray camera. Traditionally, photographic technique was used for the visualization. Visible light (photons) reduces AgCl in emulsion dispersed on a photograph paper. Then, the paper is developed and fixed as a photograph. Since EQ also reduces AgCl, the incident profile of EQ can be visualized as the photograph. This is the way that Mrs. Curie first found EQ. The technique used for visualization of EQ profiles is referred as autoradiography. As the source of EQ, T, ^{14}C, ^{32}P, and other isotopes of constituent elements of tissues and organs are used for inspection of abnormal tissues for medical purposes (see Sect. 7.6 in Chap. 7). Tritium autoradiography has been frequently applied to study hydrogen in metals which is quite important for hydrogen embrittlement, well-known cause of the sinking of the Titanic [3].

Owing to the recent development of digital photograph technology, an imaging plate or a digital camera has replaced the conventional photographic technique, which enables the visualization of EQ profiles on imaging plate or digitized film. Two examples are given in Fig. 1.7 in Chap. 1 and Fig. 5.1 in Chap. 5. The former shows the distribution of ^{41}K contained in the root vegetables, while the latter the distribution of radioactive fallout of the Fukushima accident on a vegetable leaf and also indicates that the fallout includes both β-photons and γ-particles sources.

6.5 Absorbed Dose Equivalent—Accuracy, and Assessment of Effects of EQ Exposure

In the previous sections, it is indicated that the value of absorbed dose or dose equivalent determined by a dosimeter includes some error, so as effective dose. In this section, the number of measurable digits and the accuracy in the determination of absorbed dose and dose equivalent are revisited for the assessment of the effects of EQ exposure.

6.5.1 Consideration of Absorbed Dose Equivalent (Sv) and Effective Dose to Use for the Assessment of the Effects of EQ Exposure

According to the recommendation of the International Commission on Radiation Protection (ICRP), for public protection in an emergency, reference levels for the highest planned residual doses are typically in the range of 20–100 mSv (The ICRP 2007 recommendation, Table 8 [4]) and most of countries have employed the recommendation as to their regulation level for emergency. Throughout the recommendation, the unit of Sv is used for the assessments of the effects of EQ exposure. Correspondingly, simple dosimeters commercially available display the estimated absorbed dose or dose equivalent in μSv or mSv unit.

The absorbed dose equivalent has been introduced as a common indicator (measure) to show the effect of the exposures irrespective of the kind of EQ in the study of the emergence of biological effects of EQ exposure. However, it does not seem suitable to compare the radiation effects given by the different kinds of EQ simply with the values of Sv. Because present techniques for EQ measurements easily identify the kind of EQ, distinguish different ones, and determine EQ energy, absorbed dose can be determined individually for each tissue or organ exposed to EQ. In a simple dosimeter, intensity of EQ detected as Bq is converted to absorbed dose equivalent (Sv) with the radiation weighting factor (W_R averaged for a whole body in Table 2.1 in Chap. 2). Then, the values determined as absorbed dose are converted to effective dose using the tissue weighting factors (W_T) specified for each tissue and organ (Table 2.2 in Chap. 2). Thus, for conversion of the absorbed dose to the effective dose, two weighting factors are required which would reduce the accuracy of the effective dose.

Owing to accumulation of data for the effects of EQ exposure, the data arrangement or organization using the absorbed dose or energy seems possible with distinction of exposed substances (specifying tissues or organs exposed to EQ), and identification of kind of EQ, their intensity and energy. This allows to determine the absorbed energy or absorbed dose in unit time ($Gy\ s^{-1}$, $Gy\ h^{-1}$, etc.) for the specified tissue or organ. So that it does not seem necessary to use the dose equivalent or the effective dose.

The radiation effects given by exposures to X-ray, γ-photons, and β-particles must be different from each other. Nevertheless, as given in Table 2.1 in Chap. 2, their radiation weighting factors are the same. The factors for neutrons in energy ranging from 100 to 2000 keV and particles are set to be the same as 20. In principle, the way of energy deposition caused by lighter quantum-like photons and electrons is quite different from that caused heavy quanta like neutrons and heavy ions so as their irradiation effects. Furthermore, all tissue weighting factors for organs except brain in Table 2.2 in Chap. 2 are within factor of three. As already described in Sect. 6.2.2 the measured intensity of EQ (in Bq) and absorbed dose or energy includes statistical scattering or errors. Therefore, it is not likely meaningful to give detailed figures in the weighting factors.

More significant difference appears in resilience different from organ to organ and person to person, particularly for lower absorbed dose. Therefore, discussion of the effects of EQ exposure for the lower absorbed dose, the difference of factor three is not significant but they can be expressed by logarithmic scale, i.e., the difference could be over an order of magnitude in absorbed dose equivalent or effective dose. In this respect, it is quite natural that the value of absorbed dose (Gy) and dose equivalent (Sv) for X-ray, γ-photons, and β-particles are quite close, or their radiation weighing factors are nearly unity.

Thus, it would be better to organize or summarize the effects of exposure with absorbed dose (Gy) with identification of the kind of EQ, their intensity and energy, and of the tissue or organ exposed. The use of the absorbed dose equivalent and effective dose has history and is widely employed. Nevertheless, it should be reconsidered to continue the use of the absorbed dose equivalent or effective dose in research and discussion of radiation effects.

6.5.2   Accuracy and Number of Significant Figures in EQ Measurements

As shown in Fig. 2.2 in Chap. 2 and Fig. 4.1 in Chap. 4, the absorbed dose equivalent for displaying exposure effects has expanded by nine orders of magnitude from 1 μSv to 1000 Sv. Of course, no detector can measure absorbed dose or dose equivalent with such wide ranges. For example, in scaling of length, the length of 20 cm can be measured with the accuracy of some 0.1 cm, i.e., measurable (significant) digits is three. For shorter lengths, a micrometer can be used with accuracy of some 0.01 mm. Electric microscope can measure 1 nm but cannot be used for cm scale. Similarly, EQ measurements use different methods depending on the intensity and energy of EQ. In addition, the intensity of EQ is essentially scattered owing to statistic nature of nuclear disintegration. Usually, by repeating the EQ intensity measurements, the averaged intensity and standard deviation are determined. Or approving the deviation follows the Poisson's distribution or normal distribution, the measured value is accepted as having the deviation of 2–3 times the standard deviation, or square root of the measured intensity as given in Eqs. (6.2) and (6.3), respectively. The deviation decreases with increasing the measuring time.

For higher absorbed dose exposure, e.g., absorbed dose equivalent to over 1 Sv, the difference of a few mSv can be ignored. On the other hand, for lower absorbed dose exposure, the difference of 1 mSv can result in significant difference in the appearances of the effects of EQ exposure. This is one of the causes of difficulty in discussing the effects of EQ exposure for lower absorbed dose or lower dose rate. Now for simplicity, assume the effects of the exposure are proportional to the dose rate and the number of leukocyte decreases appears over the absorbed dose equivalent of 200 mSv. Then one of ten thousand would suffer from the leukocyte decrease with the absorbed dose equivalent of 20 mSv. Is this number too many or too small? To conform this, at least 100,000 people should be inspected. If the weighting factors for determination of absorbed dose equivalent or effective dose deviated two times, the effects would not be confirmed without the inspection of additional number of people. Thus, it is extremely difficult to quantitatively predict the effects of EQ exposure for lower absorbed doses.

This argument does not mean the EQ exposure of low absorbed dose is safe. In principle, it is difficult to predict the appearance of rare events, which is essentially a statistical phenomenon. Deaths by cancers and other diseases, suicides, etc. do not happen to everyone, but in Japan, for example, more than ten thousand people are affected every year. Traffic accidents also kill similar number of people.

As mentioned in the previous chapter, the use of energy or power always accompanies some risk. The use of nuclear energy is no exception. For use of any kind of energy or power, people should assess its benefit and risk correctly and pay for the risk. Most of EQ sources on the earth that could give some influenced on people exposed to is originated from the utilization of the nuclear energy. And EQ carries some of energy produced by nuclear reactions. Since the way to deposit their energy to people is quite different depending on the kind, intensity, and energy of EQ, they should be determined to estimate absorbed energy or dose and possible effects of the EQ exposure. Still, the appearance could be modified by resilience of individuals.

References

1. https://www.miraikan.jst.go.jp/sp/case311e/home/docs/1105171618/index.html
2. J. Chadwic, C.D. Ellis, *Radiations from Radioactive Substances* (Cambridge University Press, Cambridge, 1951)
3. https://en.wikipedia.org/wiki/Sinking_of_the_Titanic
4. ICRP 2007 Recommendations of the International Commission on Radiological Protection ICRP Publication 103. Ann. ICRP 37(2–4)

Chapter 7
Utilization of EQ

Abstract Depending on how to use the energy carried by EQ (radiation), EQ can be either beneficial or hazardous. EQ is used in a variety of fields considering the nature of EQ. Although the utilization of EQ is hardly recognized in daily life, except for X-rays for medical inspection and EQ for cancer treatment, EQ is used in various fields, for example, improvement of mechanical properties of polymers and plastics by bridging caused by EQ exposure, fire alarms or smoke detectors, sterilization or disinfection, prevention of food deterioration and so on. In this chapter, a little detail on utilization of EQ is described from the aspect of how EQ energy is converted to useful ones.

Keywords Energy source · EQ utilization · Sterilization · Medical use · Radiometric dating · Radioisotope · Tracer

7.1 Introduction

Depending on how to use the energy carried by EQ (radiation), EQ can be either beneficial or hazardous. On utilization of large amount of energy, whatever its source is, there always appear side effects, often inconvenient for people or society. For example, electricity generated by the combustion of fossil fuels is quite important for daily life in the world, while the combustion produces huge amount of carbon dioxide (CO_2), one of the most concerning greenhouse gases. In recent days, an accident in boilers like explosion is seldom, but it was a big problem in early days of industrial revolution.

EQ is used in a variety of fields. Figure 7.1 is a tree showing how and in what kind of fields EQ is used. Utilization of EQ is hardly recognized in daily life, except for X-rays for medical inspection and EQ for cancer treatment [1]. In reality, however, EQ is used in various fields, for example, improvement of mechanical properties of polymers and plastics by bridging caused by EQ exposure, fire alarms or smoke detectors, sterilization or disinfection, prevention of food deterioration, and so on. One of the recent topics for EQ utilization is application of muons coming from space for observation of the inside of the melt-down reactor at Fukushima. The muons can penetrate even through the earth [2].

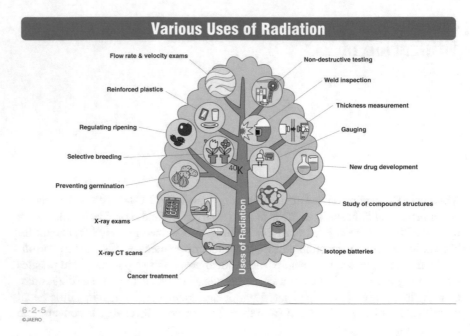

Fig. 7.1   Tree showing uses of Energy Quanta (EQ) (Radiation) in various fields (in [1], Reprinted with permission)

There are many websites showing the utilization of EQ [3–6]. In this chapter, a little detail on utilization of EQ is described from the aspect of how EQ energy is converted to useful ones.

7.2   Sterilization or Disinfection

Sterilization or disinfection with EQ exposure is based on the lethal properties of EQ which people is afraid of. Except for very high dose of EQ exposure like atomic bomb which burns human bodies to kill, chemical reactions caused by EQ inside cells destroy them or their proliferative function resulting in their death or the canceration of tissues. Although the chemical reactions caused by the EQ exposure might be different from those caused by chemicals like pesticides and disinfectants, it can be said that both are the same from the viewpoint of the death of the cell or the canceration of the tissues, i.e., they are caused by the chemical reactions in cells or tissues.

In the case of the chemicals, they contribute directly to the chemical reactions, while in the case of EQ exposure, energy deposited to some area in a cell ionizes molecules (mostly water) therein generating electrons, ions, and radicals, which in turn chemically attack important elements in the cell-like DNA. Namely, the EQ exposure causes the chemicals reactions in the cell as the secondary phenomena.

In Table 4.1 in Chap. 4 are shown lethal doses of EQ for various living beings. The lethal dose for lower living beings such as viruses and bacteria is about four orders of magnitude larger than those for higher living beings such as human beings. Therefore, sterilization or disinfection requires an intense EQ source, and usually, γ-photons and β-particles, or artificially generated electron beams or X-rays are used. Workers for that should be well monitored for their absorbed dose and protected to avoid any exposure effects.

There is a famous successful event to eliminate pest insects called Melon fly in Okinawa island in Japan with sterilization of mail Melon fly by EQ exposure. For more information, please refer to the website of the Okinawa Prefecture Pest Control in [7].

For prevention of food deterioration, EQ exposure is effective and employed worldwide with economical scale of over 2 billion $. As summarized in Table 7.1 [8], the EQ exposure is done for various foods, including spices, meats, vegetables, and so on. In Japan, the EQ exposure is allowed only for onions and potatoes, while many countries allow using the EQ exposure for various foods. Commercially available spices are most frequently EQ exposed ones. However, there is no need to worry too much about irradiation of foods. As mentioned above, foods never become radioactive, or radiation sources are never transferred to foods by the EQ exposure (US Food & Drug Administration gives overview of food irradiation in [9]).

7.3 Medical Purposes

Perspective inside imaging is a simple technique and widely used including medical purposes. Just put an object between an EQ source (X-ray, γ-photons, and electron- and ion beams) and a detector. EQ transmit the object with some attenuation depending on the density profile of the object and deposit their remaining energy on the detector to give density profile of the object. As the EQ source either X-rays, β-particles, or electron beams, γ-photons and ion beams can be used, while a fluorescent screen or imaging plate is used as the detector. In the former, the energy of EQ injected is transformed into light (fluorescence) to make a photograph, and in the latter, the energy for the electron excitation in imaging plate is converted to stimulated fluorescence afterward to give photograph. For imaging, nowadays, a digital camera is employed instead of conventional photograph techniques.

X-ray fluoroscopy is used most often for medical purposes and inspection of dieses in organs. Another important use of EQ is for cancer treatment. Its principle is to deposit energy of EQ focusing on a canceration tissue and kill cancer cells in it. Therefore, the kind of EQ is selected to give maxim energy to the canceration tissue as summarized in Table. 7.2.

As for cancer treatment with EQ, there is a detailed description of "types and methods of radiation therapy" on the website of the Cancer Information Service of

Table 7.1 Amounts of irradiated foods in the world

Country		Processed amount		Irradiated foods
		2005	2010	
1	China	146,000	266,000	Garlic, Spices, Cereals, Meet, Others
2	USA	92,000	103,000[a]	Meet, Fruits, Spices
3	Ukraine	70,000	?	Wheat, Barley
4	Brazil	23,000	?	Spices, Dried herbs, Fruit
5	South Africa	18,185	?	Spices, Others
6	Vietnam	14,200	66,000	Frozen seafoods
7	Japan	8090	6246	Potato, Onion
8	Belgium	7279	5840	Frog legs, Chickens, Shrimps
9	Korea	5394	300	Dried agricultural products
10	Indonesia	4011	6923	Cocoa, Frozen seafoods, Spices
11	Netherland	3299	1539	Spices, Dried vegetable, Chickens
12	France	3111	1024	Chickens, Frog legs, Spices
13	Thailand	3000	1484[b]	Spices, Fermented sausage, Fruits
14	India	1600	2100[b]	Spices, Dried vegetable, Fruits,
15	Canada	14,00	?	Spices
16	Israel	1300	?	Spices
17	Mexico	–	10,218	Fruit (Guava), Others
	Others			
	Total	404,804	47,461[c]	
			(577,000)[d]	

[a] 15,000 tons of Fruits, Vegetable do not include Fruits imported from Mexico
[b] Excluding those processed by private company
[c] Excluding those processed in Ukraine, Brazil, South Africa, Canada, and Israel
[d] Estimated world total employed the processed amounts in 2005 for Ukraine, Brazil, South Africa, Canada, and Israel
[Source, Kume, Tamikazu, Current status of food irradiation in the world, Shokuhinn Shosa, Vol.49(2014)115. (In Japanese)]

the National Cancer Research Center [10]. Tables 7.2 and 7.3 summarize EQ sources and irradiation methods used in medical therapy. Various EQ sources are used. They are γ-photons, X-rays, and particle beams such as proton beams, deuteron beams, heavy particle beams, and neutrons.

Different from visible right, γ-photon and X-rays are hardly refracted in materials and are difficult to be focused with using a lens. If the refraction and focusing of γ-photons were possible, its safety handling would be easier. The different characters between the visible light and X-rays and γ-photons are essential owing to the shorter wavelength of the latter. To make their refraction possible, extremely high density materials are required, which is not existing on the earth. Therefore, for radiation therapy with using external exposure of γ-photon and X-rays, the exposure to healthy

Table 7.2 Kinds of EQ and irradiation methods used in medical therapy

	Kind of EQ Used		Method
External radiation	Electron beam		Energy particle beam therapy
	X-ray		
			Three-dimensional conformation therapy
			Intensity Modulated Radiation Therapy (IMRT)
			Image-Guided Radiotherapy; (IGRT)
			Stereotactic RadioTherapy (SRT)
	γ-ray		Stereotactic RadioSurgery (SRS)
	Particle beam		Proton beam Therapy
			Deuteron beam Therapy
			Boron Neutron Capture Therapy (BMCT)
Internal radiation	Brachytherapy	α-particle	Implant radiation
		β-particle	
		γ-photon	
		γ-photon	Intracavitary irradiation
	Radiopharmaceutical therapy	α-particle	Internal radiotherapy
		β-particle	
		γ-ray	

Table 7.3 Charged particle beam radiation therapy

	Kind of particle beam	Method	
External radiation	Proton beam	Ion beam radiation therapy	Proton beam radiation therapy
	Deuteron beam		Deuteron beam radiation therapy
	Neutron	(BNCT: Boron Neutron Capture Therapy)	

areas near canceration tissue is hardly avoided. For external irradiation, a suitable method to irradiate a high dose only to the affected area is selected. For internal exposures, a method to bring an encapsulated EQ source (mostly a radioisotope) into the body or directly implant or deliver the radioisotope to the canceration tissue.

Charged particle beam radiation therapy is a rapidly advancing technique in recent years (See Table 7.3). Particle beams, especially using protons (p or H^+), or deuteron (D^+) beams can be focused to a canceration tissue. As described in Chap. 4, charged particles injected in an object deposit energy concentratedly near their projected range with nuclear stopping. Hence, choosing the incident energy of the beam, its projected range can be set at the location of the canceration tissue. Neutron is also used as radiation therapy. Since neutrons have no charge, the probability of colliding with atoms and electrons is not large. However, a special nucleus like boron-10 (^{10}B) shows a very high collision cross-section with neutron resulting in nuclear transmutation as,

$$^{10}B + n \rightarrow\ ^7 Li +\ ^4 He. \tag{7.1}$$

Li and He produced by the reaction carry energy released and kill cells nearby. First making a chemical compound including ^{10}B, which can be easily trapped in tissues or organs including canceration cells, then it can be delivered by injection. Afterward, neutron is exposed from the outside to make the nuclear reaction.

7.4   Utilization of EQ Energy

EQ, of course, can be used as energy source. A nuclear reactor described in Sect. 3.3 in Chap. 3 is exactly the one that converts the energy of EQ produced by nuclear fission into heat to generate electricity.

The energy stored in radioisotopes (RIs) can be also used. Different from fossil fuels which require oxygen to combust, RIs continue to release energy as the emission of EQ without oxygen. A nuclear battery is used as electric power supply in satellites, in which all emitted energy of EQ from RI is converted to heat and then the heat is converted to electricity with thermoelectric elements. For RI, ^{238}Pu is often used which decays emitting α-particles with the half-life of 87.7 years with output power of $540\,W\,kg^{-1}$. Owing to its very short penetration length of the α-particles, most of their energy is deposited within very short distance and converted to heat, accordingly little shield is required. Thus, the nuclear battery is maintenance-free and has a long-life.

Solar cells are driven by visible light, so as the exposure of X-rays and γ-photon to the solar cell also generates electricity. However, the conversion efficiency of their energy to the electricity is very poor, because most of them permeate through the cell and only tiny part of their energy is deposited in the cell. In addition, the exposure induces radiation damage in the cell made of semiconductors and significantly degrades the conversion efficiency. In this respect, β-particle is better. Also, β-particles can be collected directly as an electric current and used as a β-voltaic nuclear battery.

Figure 7.2 describes a parallel plate capacitor-type nuclear battery, which is still under development but is useful to understand how energy of EQ (γ-photon) is converted to electricity. As described in Chap. 4, when EQ is injected into a material, their energy is deposited through electron excitation (ionization) and nuclear collisions. In case of γ-photons, most of their energy is deposited through the former process and many excited electrons appear around their trajectories (referred as

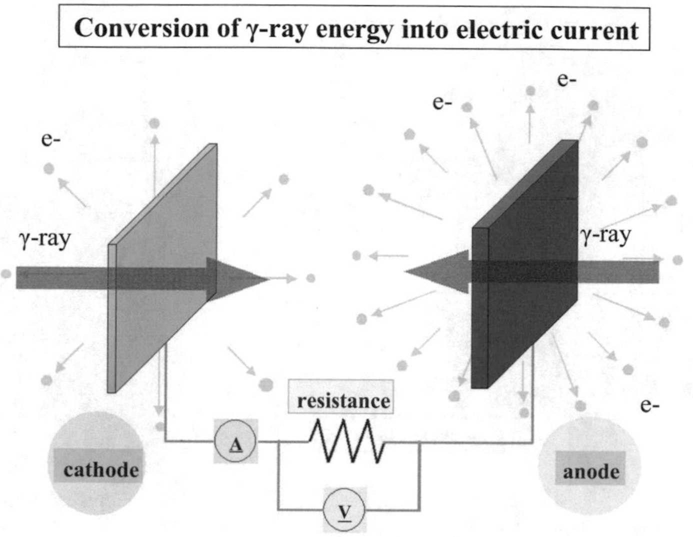

Conversion of γ-ray energy into electric current

Fig. 7.2 Conceptual view of a parallel plate-type radiation battery. When two metal plates with the different thicknesses are exposed to γ-rays, the electron emission from thinner plate is larger. On the other hand, for two metal plates with the same thickness, but different in atomic numbers, the emission of electrons from the plate with larger atomic number is larger. When the two plates are connected through a load (resistance), the current flows and power is deposited into the load

secondary electrons). The secondary electrons succeedingly collide with atoms and electrons around them creating tertiary electrons. This ionization or electron excitation processes continue until all generated electrons lose energy not enough to ionize atoms in the material.

If the material is a thin plate-shaped, certain part of the generated electrons escapes from the surface of the plate. If two metal plates made of different metals are put parallelly such as parallel plate capacitors, different numbers of emitted electrons between the two plates bring difference in their electric potentials. If the energy of incident β-particles is 1 MeV (10^6 eV), ideally it could generate 10^5 electrons with 10 eV of energy. In reality, it is not possible to generate such large number of electrons in a material. Most of generated electrons are recombined with ionized atoms releasing the energy as heat. This is the principle of the shielding of EQ.

If the plate is thinner than the moving distance or projected range of the incident EQ, the generated electrons can get out of the materials, because the heavy material includes more electrons to be excited and EQ incident in it lose shorter depth as shown in Fig. 7.3 [11]. The heavier the material, the larger is the number of electrons generated. On the other hand, if the plate is thick, secondary electrons are less likely to come out, so a potential difference occurs between the two electrode plates because there is a difference in the number of electrons emitted between the two plates with one plate made of heavier material and the other made of lighter material (alternatively two made of the same metal but different in thickness). When the

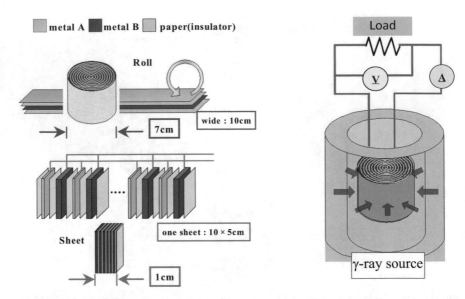

Fig. 7.3 Conceptual design for self-shielding radiation batteries using γ-source. Power production and shielding are simultaneously done by setting many parallel plates in front of the source or surrounding the source to make concentric cylinder [11]

parallel plates are connected through a load, power is obtained. This is the concept of the parallel plate nuclear battery as shown in Fig. 7.3. In the right figure is given a self-shielding γ-photon battery in which γ-photon source is set at the center, and surrounded by many bipolar plates with insulators in between. Since multi-plates work as shield, no shield is required.

7.5   Radiometric Dating (^{14}C dating)

EQ are used in a variety of fields. As already mentioned, perspective views using X-rays and γ-photons allow to see the inside of objects and are applied not only for medical inspections but also for nondestructive inspection of variable arts and treasures. In criminal investigations, various EQ are used in the field of analysis as part of legal science and technology and are used for inspection of carry-on baggage for explosives and drugs, and micro or trace analysis including radiation analysis.

In the following, particular focus is given to ^{14}C dating. ^{14}C is a radioactive isotope of carbon present in nature. The historical changes in carbon dioxide (CO_2) concentration in the air, are extremely important in discussing recent global warming. Figure 7.4 [12] shows the historical change of CO_2 concentrations in the atmosphere since 1000 A.D. The CO_2 concentrations in the atmosphere have been kept at nearly 280 ppm in the past, while recently they have risen to near 400 ppm, which is believed to be the cause of global warming. As described in Chap. 2, CO_2 in the

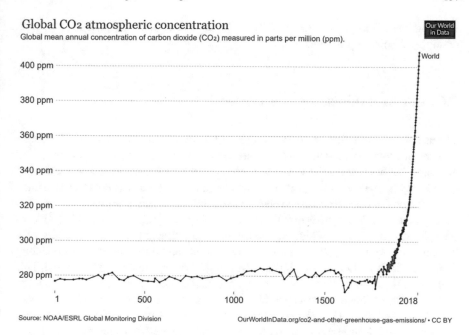

Fig. 7.4 Changes in CO_2 concentrations in the atmosphere from AC 1000 to 2000 (in [12], Open to public)

atmosphere absorbs radiation (infrared rays) from the earth and then releases longer wavelengths of infrared light to both upside and downside. Hence the increase of the CO_2 concentration in the atmosphere will increase the energy return to the ground surface and lead to global warming.

By the way, how is determined the CO_2 concentration in the atmosphere more than 1000 years ago? As described in Chap. 3, ^{14}C is generated with the collision of nitrogen in the atmosphere and neutrons included in cosmic rays by 7.5 kg of ^{14}C per year, while it decays to ^{14}N emitting β-electrons with a half-life of 5730 years. Since the production and decay of ^{14}N are balanced in the atmosphere, the concentration ratio of stable isotope of ^{12}C and radioisotope of ^{14}C in normal water and air is kept constant. In living beings, the ratio is kept to the same as that in the atmosphere when they are alive. While after their death, their metabolism stops, i.e., ^{14}C in the dead body cannot be exchanged with that in the atmosphere, and accordingly, it decays following the lifetime or the concentration of ^{14}C in the dead body simply decreases. Using this phenomenon, one can measure the ^{14}C concentrations contained in unearthed wood, paper, and so on, to see how old they have been since they died. On high mountain areas and Antarctica, layers of ice have accumulated for a long time and trapped air as bubbles in which the concentration of ^{14}C simply decreased. Taking these bubbles and measuring both CO_2 concentration and the radio of ^{12}C and ^{14}C, one can determine both how high the CO_2 concentration in the atmosphere was and how long years have passed after they were trapped in the ice. Thus Fig. 7.4

was obtained. Without the generation of ^{14}C by the cosmic rays (neutrons), such dating could not be possible.

7.6   Use of Radioisotope as Tracers

The use of ^{14}C for dating in the previous section was not possible without the development of radiation measurements. Detection of EQ is so sensitive that very tiny amount of radioisotopes, less than ppb levels, can be easily detected. A method of examining the material transport in tissues, organs, and a body is developed using a radioisotope (RI) as a tracer and referred to as the tracer technology.

Often used RI are tritium (T or ^{3}H), carbon-14(^{14}C), fluorine-17 (^{17}F), or phosphorus-32 (^{32}P) of which stable isotopes are fundamental constituents of a human body. Metabolic abnormalities and lesions can be examined, with injecting some chemicals including such RI, and profiling RI using an imaging technique. This tracer technique with RI is used not only for medical purposes but also for quality improvement of flowers and vegetables. Owing to easy detection of RI, its amount used is so small that its irradiation effects could be quite small. In addition, the decay times of these RIs are rather short, and their biological half-lives are also short (See Table 5.1 in Chap. 5).

For use of the tracer technique, the progress of visualization technology for distribution of RI in materials has played a quite important role. A good example is shown in Fig. 7.5 [13] which is a 3D color map from a PET/CT scan using ^{17}F for a mouse bearing an *E. coli* infection on the left leg. Moreover, the technique is totally relied on the fact that EQ carries energy, otherwise, it cannot be realized.

Fig. 7.5 3D color map from a PET/CT scan with using ^{17}F for a mouse bearing an *E. coli* infection on the left leg. The 3D projection also shows tracer accumulated in the liver, kidneys, and bladder. (Reprinted from [13] by Gowrishankar G, Namavari M, Jouannot E, Hoehne A, Reeves R, Hardy J, Gambhir S, licensed under the Creative Commons Attribution 4.0 International license)

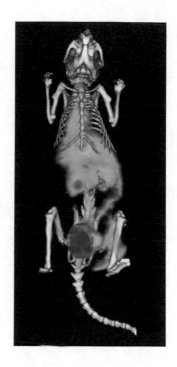

References

1. https://www.ene100.jp/www/wp-content/uploads/zumen/e6-2-5.pdf
2. H. Fuji, K. Hara, K. Hayashi, et al., *Investigation of the Unit-1 nuclear reactor of Fukushima Daiichi by cosmic muon radiograph*. Prog. Theor. Exp. Phys. 043C02 (14 pages) (2020). https://doi.org/10.1093/ptep/ptaa027
3. http://criepi.denken.or.jp/jp/rsc/knowledge/index.html(in Japanese)
4. https://www.nrc.gov/about-nrc/radiation/around-us/uses-radiation.html
5. https://www.mirion.com/learning-center/radiation-safety-basics/uses-of-radiation
6. https://www.radiationanswers.org/radiation-sources-uses.html
7. http://www.pref.okinawa.jp/mibae/website
8. T. Kume, *Current status of food irradiation in the world*. Shokuhinn Shosa **49115** (2014) (In Japanese)
9. http://www.fda.gov/food/irradiation-food-packaging/overview-irradiation-food-and-packaging
10. http://ganjoho.jp/public/dia_tre/treatment/radiotherapy/rt_03.html
11. T. Yoshida et al., *An attempt to direct conversion of gamma-ray into electricity*. Nucl. Sci. Eng. **150**, 362–367 (2005)
12. https://ourworldindata.org/grapher/global-co-concentration-ppm
13. https://commons.wikimedia.org/wiki/File:Investigation-of-6-18F-Fluoromaltose-as-a-Novel-PET-Tracer-for-Imaging-Bacterial-Infection-pone.0107951.s002.ogv

Chapter 8
Energy and History of the Earth

Abstract Discussed are energy balance of the earth between input (mostly from sun) and output (as radiation of electromagnetic waves in infrared region) and its historical changes. In most of times of Earth's 4.6-billion-year history, there was no oxygen in the atmosphere, and due to the influence of EQ from the sun, living beings hardly appeared, or were not allowed to live. At the beginning of the birth of life around 3.5 billion years ago, the temperature of the earth was too high to survive for higher living beings. The appearance of human beings was only 100 million years ago and only two centuries have passed since human beings were able to use fossil fuels stored in the Carboniferous period as an energy source. It is only recent 100 years that the human beings have been combusting large amounts of fossil fuels to get energy and released huge amounts of CO_2 into the atmosphere. Considering that the ancient atmosphere mostly consisted of CO_2, the recent increase of CO_2 in the atmosphere is in some sense returning to the ancient age when the temperature of the earth was much higher than now.

Keywords Energy balance · Earth · Fossil fuel · History of the earth · Influence of EQ · Sun

8.1 Introduction

In most of times of Earth's 4.6-billion-year history, there was no oxygen in the atmosphere, and due to the influence of EQ from the sun, living beings hardly appeared, or were not allowed to live. At the beginning of the birth of life around 3.5 billion years ago, the temperature of the earth was too high to survive for higher living beings. The appearance of human beings was only 100 million years ago. Furthermore, only two centuries have passed since human beings were able to use fossil fuels as an energy source. Combustion of the fossil fuels has produced so many CO_2 to influence the atmosphere as a greenhouse gas. Considering that the ancient atmosphere mostly consisted of CO_2, the recent increase of CO_2 in the atmosphere is in some sense returning to the ancient age where the temperature of the earth was much higher than now. In the Carboniferous period, around a few hundred million years ago, CO_2 in

© Kyushu University Press 2022
T. Tanabe, *Radiation: An Energy Carrier*,
https://doi.org/10.1007/978-981-19-1957-2_8

the atmosphere started to decrease owing to photosynthesis of plants which transformed CO_2 into organic matters (carbohydrates) and preserved it in the ground. Human beings mine them, combust and return them to the atmosphere as CO_2. It is only recent 100 years that the human beings have been combusting large amounts of fossil fuels to get energy and released huge amounts of CO_2 in the atmosphere. In this chapter, discussed is energy balance of the earth between input (mostly from sun) and output (as radiation of electromagnetic waves in infrared region) and its historical changes.

8.2 Changes in Global Environment

As shown in Fig. 1.4 in Chap. 1, the global environment is maintained by the balance between energy input from the sun and output as radiation from the earth. Occasionally there appeared power imbalance caused by volcanic explosions, meteorite impacts, and so on, resulting in short periods of ice ages or warming. Compared to these natural climate variations, recent global warming, i.e., the increase of the average temperature over the last 20–30 years, is remarkable and is attributed to the changes in energy balance caused by the increase of greenhouse gases, in particular CO_2, in the atmosphere. In addition, local heat island events are appreciable in megacities, which are caused by waste heat as a result of combustion of fossil fuels.

Thus, it becomes clear that energy input from the sun and the changes in the global environment of the earth are closely connected. Figure 8.1 shows the history of the earth in billions of years. The figure also shows changes in the input power from the sun, the intensity of UV at the ground surface, and partial pressures of nitrogen (N_2), oxygen (O_2), and CO_2.in the atmosphere. In long time scale, the brightness of the sun is increasing, while the intensity of ultraviolet light on the surface of the earth was decreasing. The CO_2 concentration in the atmosphere has been also decreasing while O_2 concentration (or partial pressure) increasing. N_2 concentration has stayed rather constant. About a billion years after the birth of earth, or about 4 billion years ago, the atmosphere was mainly composed of N_2 and CO_2. The surface of the earth was composed of inorganic materials, and no living beings existed. It is said that the lives, which were organic matters that can self-proliferate, were born about 3.5 billion years ago. Energy is required to synthesize organic matters (carbohydrates) in an environment containing only water (H_2O) and CO_2, or to proliferate living beings. The sun gave enough energy to the surface of the earth. At the same time, however, it gave dangerous EQ to the living beings. Hence the living beings could survive in sea because the seawater absorbed the higher energy EQ from the sun. Some organisms existed getting energy from hydrothermal deposits (underwater volcanoes) in the water. (Of course, some still exist today.)

Although the details of how organic matter and living beings appeared from inorganic materials are not well understood, about 2.5 billion years ago, appeared are some living beings like algae that had the ability of photosynthesis reducing CO_2 with H_2O using the solar energy-producing carbohydrates. At that time, although

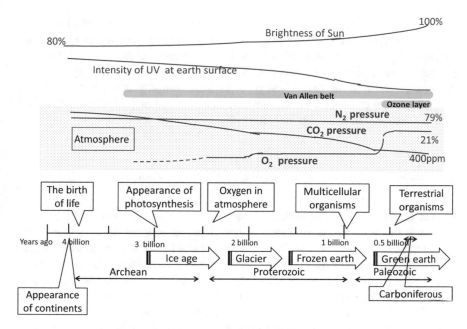

Fig. 8.1 History and changes in atmosphere of the earth

luminosity or brightness of the sun was a little less compared with the present sun, the intensity of ultraviolet light was much stronger than now. Therefore, the first algae or some living beings having the ability to make photosynthesis appeared and thrived in the sea. Since O_2 released by the photosynthesis does not dissolve in water, the concentration of O_2 in the atmosphere had being increased. As the O_2 concentration in the air increased, the intensity of the ultraviolet light reaching the surface gradually decreased due to the absorption by O_2 (some are used for generation of ozone (O_3) layers), and the living beings moved on the land and have thrived. And even though the energy emitted by the sun itself is 20% larger than it was at the birth of the Earth, fortunately, the UV light reaching the surface has decreased.

About 0.5–1 billion years ago, with increase of terrestrial plants, CO_2 levels significantly decreased, while O_2 increased. Accordingly, the intensity of the ultraviolet light further decreased and the surface of the earth became a place where living beings were safe to live, and the diversification of living beings was promoted, i.e., the promotion of different types of living beings, like dinosaurs, insect, animals, etc. progressed. Only for one-tenth (1/10) period, or 500 million years of the earth history, the surface of the Earth has been safe for living beings owing to the shielding of dangerous ultraviolet or shorter-wavelength light by the atmosphere. It should be noted that until now no planets other than the earth with similar atmosphere and living beings have been found in the universe. The earth is the exceptional place.

Looking back at Fig. 1.4 in Chap. 1, one can see that only 0.23% of the total input power from the sun is used for photosynthesis and stored from the aspect of energy balance. Such energy had been stored as fossil fuels (colas, gases, and oils) during carboniferous period (from 0.36 to 0.29 billion years ago) as shown in Fig. 8.1 and the present people were benefiting from them. In other words, the fossil fuels are just solar energy given by the sun and stored in old days. Modern society consumes energy equivalent to the stored energy for about a million years in just one year. If the current pace of fossil fuel consumption will continue in the future, stored fossil fuels for future is only for few hundred years.

8.3 Development and Evolution of Living Beings

No one denies the influence of EQ exposure on occurrence and evolution like diversification, or higher forms of living beings. The evolution of living beings is not something that has been made with necessity, but that has been occasionally selected to be fitted with the environment from those given by mutation in nature. The question is why mutation evolution appears. The EQ exposure is one of the possible causes. The heavy exposure would result in the death of cells or tissues if the function of keeping cells alive in RNA and DNA was damaged. However, some other changes in RNA and DNA owing to the induced chemical reaction caused by the EQ exposure would introduce some change in characters of tissues, organs, and a body leading to so-called mutations. The birth of the first life would have been the generation of some kind of organic matter that can be self-proliferating. The birth of photosynthesis-capable lives such as algae or plants is the appearance of a special function in the cells that can reduce CO_2 with H_2O to generate carbohydrates. Specialty of the plants is to use the energy given by sunlight for the photosynthesis. Early plants that initiated photosynthesis may have used higher energy photons than the current visible light. Because in early years, the intensity of ultraviolet light was stronger than that is now, and higher energy photons is more advantageous to induce chemical reactions. It seems quite natural to appear animals that gain energy required for sustaining their lives by eating the plants and burning them through metabolism. The dead bodies of algae and plants have been converted to be fossil fuels and the carcasses of animals to be limestones, or $CaCO_3$ of which Ca is the main component of their bones and shells.

In any way, almost of all living beings on the earth are getting energy for their survival from the sunlight. Energy emissions from the sun have increased by about 20% from the birth of the earth to today, as shown in Fig. 8.1. Nevertheless, the energy received on the earth's surface has not increased due to the absorption of atmosphere, and the intensity of the ultraviolet light dangerous for the life beings is significantly reduced. Therefore, the appearance of mutation and/or canceration of higher organisms would be less in the present earth, or the present earth is much safer to live for the higher living beings. The transformation of information in nerve systems is done as the transformation of energy or electrical signals, which could be disturbed by EQ, such as ultraviolet light and soft X-rays. The ultraviolet light and soft X-rays have some additional effects other than the induction of the chemical reactions in cells described above. For example, it is well known that solar flares disrupt electromagnetic waves on the ground. If a huge flare occurs, hazardous EQ not only for the human body but also for communication systems reaches the ground.

Whether one likes it or not, EQ from the sun always supports human beings, forces some change, and gives some health risks. Most people think that radiation (EQ) is limited to high-energy quanta. Since the radiation (EQ) carries energy, even low-energy EQ can give influence when a human body is exposed to high intensity of lower energy EQ. Sunburn is a typical example. The deposited energy which is given by the product of energy carried by each quantum and its intensity is transformed finally to heat. As mentioned several times, the appearance of the EQ exposure effects varies appreciably depending on how high the energy and intensity of EQ or absorbed dose are.

One of most concerned is the impact the EQ exposure of human beings on later generations. We cannot make experiments to make this clear to humanity. However, the evolutionary process of living beings in history gives us a suggestion for this. The EQ exposure can cause some mutation in an individual exposed to EQ. However, the evolutionary process investigated with experiments and experiences of the EQ exposure to insects and plants shows that the mutant ones cannot survive if they do not fit the surrounding, i.e. natural selection works. As one of the living beings, human beings have been also exposed to EQ in their long history and could have been subjected to the processes of mutation and natural selection. As the result, only homo sapience remains as human beings on the present earth. In addition, the human beings might have acquired a certain degree of resistance or resilience to EQ exposure or it may be necessary to be continuously exposed to a certain dose of EQ in order to maintain the resilience. In reality, although not a small number of people were exposed to heavy radiation given by the atomic bombs in Hiroshima and Nagasaki and suffered from disease, no significant impacts have appeared in later generations. It should be mentioned that some are still suffering from the exposure. Therefore, it is not appropriate for this discussion to apply to individual examples or changes in only two or three generations.

It is well known that pollution and chemical damage are caused by some organic metals, pesticides, insecticides, etc., and could result in diseases that affect genes. Some chemicals that are taken in cells can kill the cells or cause some changes in DNA and/or RNA, causing mutation. There is no difference in biochemical reactions caused by the chemicals and the EQ exposure because the energy deposited by EQ is converted to chemical energy in the cells.

Chapter 9
Final Chapter—Energy Resources and Radiation

Abstract All living beings can be existing on the earth with using energy supplied from the sun. Although the sun radiates energy or power as various types of EQ, the atmosphere of the earth shields higher energy (dangerous) ones. Accordingly, the energy given to the ground surface is mostly as electromagnetic waves of UV, visible and infrared lights, so that the earth is not dangerous for living beings including human beings. Any energy sources, whatever they are, accompany some risks when they are used and used energy is finally converted to or wasted as heat. This is also true for nuclear energy or EQ. Because of high energy carried by EQ, EQ is dangerous and scary. Nevertheless, EQ are existing in nature and can be used as energy source under ensuring safety.

Keywords Energy source · Energy conversion · Nuclear energy · Risk · Safety · Solar energy

9.1 Introduction

All "living beings" can be existing on the earth with using energy supplied from the sun. Although the sun radiates energy or power as various types of EQ, the atmosphere of the earth shields higher energy (dangerous) ones. Accordingly, the energy given to the ground surface is mostly as electromagnetic waves of UV, visible and infrared lights, so that the earth is not dangerous for living beings including human beings. Most of energy or power is given by the sun is once absorbed in the earth and emitted from the earth as infrared radiation. Small portion of the energy is stored in plants after conversion with photosynthesis. There are no other stars/planets where living beings are existing.

Any energy sources, whatever they are, accompany some risks when they are used and used energy is finally converted to or wasted as heat. This is also true for nuclear energy or EQ. Because of high energy carried by EQ, EQ is dangerous and scary. Nevertheless, EQ are existing in nature and can be used as energy source under ensuring safety.

© Kyushu University Press 2022
T. Tanabe, *Radiation: An Energy Carrier*,
https://doi.org/10.1007/978-981-19-1957-2_9

9.2   Energy Sources

As mentioned in Chap. 8, EQ from the sun is the source of our energy. Fortunately, both the sun herself and the earth's atmosphere shield dangerous EQ for living beings, and delivered energy is enough to live on. We understand that EQ carries energy, and we can control it to some extent. However, we cannot change the energy of EQ or stop EQ at will. Therefore, to avoid the exposure to the "scary" EQ (radiation)" it is the only way to keep away from or to shield its source. Nevertheless, it is not necessary to set the allowable absorbed dose equivalent low more than necessary. All living beings on the earth have been exposed to some EQ since ancient times, and seem to acquire the ability to tolerate slight EQ exposure. One should remind that when growing plants, they often grow more in a some-harsh environment than in a normal environment, suggesting that the plants have abilities to correspond to harder surroundings to live. From this point of view, the appearance of hormesis on EQ exposure in living beings might be reasonable. Of course, overconfidence should be avoided, but it is not necessary to worry about the EQ exposure of the absorbed dose equivalent to about 10 times higher than the natural dose.

However, there is no threshold level in the absorbed dose equivalent to ensure the absolute safety. Somewhere in nature, there could be some special events that no one has experienced or expected. For example, unknown EQ from the cosmos could be the cause of unpredictable accidents. Powerful earthquakes and typhoons are local emissions of energy/power on the earth. Moreover, the appearance of the lethal disease by the virus and the bacterium can never be zero. The only way we can do this is to "be afraid", prepare for accidents, and take the right action.

Energy use always accompanies some danger/risk. The more the energy used is, the larger the impact of the accident. The Great East Japan Earthquake is caused by the release of enormous amounts of stored energy in the earth's crust. Aftershocks are releases of the residual energy. The Fukushima nuclear accident triggered by the earthquake is also a very large emission of energy that remained in the nuclear reactors. The radioactive materials including FPs released at the accident were dispersed onto surrounding areas and continue to release the stored energy as EQ.

When the intensity of EQ from radioactive materials is high, they can be used as an energy source and are actually used. However, when the intensity is low, it is hard to use effectively. Because conversion of the energy released from low-density EQ sources to useful energy requires more energy than that the source originally has. Ironically, the existence of high-density EQ sources, which are "scary radiation", allows using their energy after converting to useful ones under safely controlled conditions as realized in nuclear power plants.

From the above-mentioned view, employing nuclear power would be an important option for a long-term energy security, although it seems quite hard to change recent trend to hesitate using nuclear energy after experiencing the nuclear accidents. As a researcher who has been involved in nuclear power development, the author's view on energy security is given below.

9.3   There's no Energy Source to Use for Free

There is no doubt that using renewable energy should be promoted, but it should be reminded that how much energy is needed to develop the infrastructure and who takes the risk for the energy conversion from various forms of solar energy to useful energy for us. The argument should be also on utilization of fossil resources. In general, economic scale-like GDP (General Domestic Product) of a country is nearly proportional to its energy consumption. Moreover, the prices of goods are strongly correlated to the energy used to produce them. "Cost" of goods is, in some sense, an indicator of how much energy was used to produce the goods. The high labor cost means to hire a person who can manage enough energy to produce and trade goods, to hire persons, and so on.

The high cost of power generation by solar cells and wind power generators is due to not only the manufacturing cost of their equipment but also the high costs of installation and maintenance of their infrastructures and transmission/connection to power grids, as well as low availability (no operation at night or no wind). At present, the cost of renewable energy is higher than that of the fossil energy. However, it is difficult to judge whether a certain energy source should be employed or not simply by cost base. Productors or promoters of the renewable energy systems claim that the cost should go down because of the future technology development and possible mass production. While consumers could have different opinions to accept presently cheaper ones.

Generally, whatever the sources are, the cheaper ones would be better. In this respect, coal is presently the cheapest among fossil fuels. However, combustion of more coal would enhance global warming. Now consideration of environmental issues becomes mandatory in any activities using energy. Conservation and saving of fissile fuels are critical. Nevertheless, as experienced after the Fukushima nuclear power plant accident, too much energy saving, which appeared as shortage of electricity at that time, had led to stagnation in economic activity. Reduction of energy consumption could shrink economic activity and, in turn, the vitality of the country.

In developed countries such as EU, Japan, and US, the labor costs occupy major part of the total cost of goods and services. However, when new production, such as solar cell production, begins, there appear preferable phenomena such as an increase in employment, which contribute to the increase of GNP. In long run, even if the cost was high (or cheap), other factors like a national policy would force to start a new project (or to stop the presently working project), e.g., in a few countries, it becomes a national policy to stop nuclear power plants despite its cheaper cost. Under the renewable energy bill passed by former Prime Minister Naoto Kan, in Japan, electric power companies are required to employ renewable energy as much as possible and extra cost needed to do this is charged to consumers. The bill has clearly promoted the use of electricity generated by solar cells. Here, the political decision was given priority over the cost. However, this overcharge on the consumers cannot be ignored these days, and some electric power companies hesitate to employ further use of the solar cells. Of course, a nuclear plant should be much safer and hence its cost would be higher than present one. Thus, the long-term cost estimation becomes harder and harder to determine, and a question "who will be responsible for the cost" is a matter of the country or the world.

9.4 Fossil Fuels Are Originally Solar Energy

As mentioned in Chap. 8, fossil fuels are ancient solar energy stored in the ground in the form of coal (C), and oil/gas (hydrocarbons) over millions of years. Around 200 years ago, people have started to use them. Nowadays the stored energy for about a million years is consumed in about a year. If we continue to consume fossil fuels in the present pace, remaining fossil fuels would be for a few hundred years, because the period when the solar energy had been stored is less than a billion years (see. Fig. 8.1 in Chap. 8). In some sense, consuming (combusting) fossil fuels as an energy source is the pass to return the present earth to the ancient earth in which atmosphere was mostly consisting of CO_2 and N_2. In present days, CO_2 can be recovered and stored to prevent global warming, but it will require more energy or high cost.

Now, renewable energy means solar energy given for one to a few years and is used within the same years without influencing energy balance of the earth. Hence as long as the relationship between the sun and the earth continues to be as it is (to date, the sun has burned about one-third of its hydrogen fuel by nuclear fusion reactions, and still two-thirds remain.), there is no need to worry about the depletion of solar energy for the next few billion years. It is a matter of controversy that how to use renewable energy among solar cells, wind power generators, and photosynthesis for biofuel and foods. The conversion efficiency of the solar cells is around 20% in highest case, while that of the photosynthesis is rather poor but the photosynthesis contributes to the reduction of CO_2 and is essential for secure foods. From the point of view of energy use, the utilization of the solar energy as biofuel or food is basically the same. The difference among them is whether human beings use them externally as an energy source or take them directly as food energy.

Here, it should be reminded that the solar energy is energy generated by nuclear fusion using hydrogen as a fuel. The energy generated by the nuclear fusion, like the energy generated by nuclear fission, is released as EQ. However, the dangerous part of EQ does not reach the ground surface of the earth owing to the shielding of the sun's interior and in the atmosphere of the earth, just like shielding by water, pressure vessels, containment vessels, etc. in nuclear reactors. Although some radioisotopes that remained in the earth's interior emit EQ, the crust shields them and their energy is converted as the decay heat, which retards the earth's cooling. All these circumstances allow living beings to exist on the earth by luck. There is no other planet like the earth.

In the moon without atmosphere, where the energy supply per unit area from the sun is nearly the same as that of the earth, human beings cannot live without a spacesuit to supply oxygen and shield dangerous EQ, or cosmic rays. Again, the circumstances of the earth are just the place where living beings survive with energy generated by nuclear reactions and converted to safer ones. It is a system in which dangerous EQ does not reach but is converted to safer one. No other planet other than the earth exists in universe where living beings exist.

In order to continue prosperities of human beings on the earth long, we have to use energy of EQ produced by nuclear reactions, which is the energy source of the universe, with energy conversion from dangerous EQ to safer ones. Of course, it would be possible to live with solar energy alone. Considering Edo ear in Japan (from 1600 to 1868) where no fossil fuels were used with population nearly constant (25–30 million), number of people who survive with solar energy alone would be around one-fourth of the present Japanese population, around 120 million. Of course, with increasing the energy conversion efficiency of the solar cells, wind power generation, and photosynthesis, a little more people can live, but at most twice as much, that is, about half of what it is today.

Living standards of under-developing countries, whose total energy consumption per capita or GNP is around 1/10 of the developed countries, are like that in the developed countries several decades ago. However, people in the developed countries, who have gotten used to the present lifestyle using lots of energy, do not seem to be able to return to the life without fossil energy more than 200 years ago, before industrial revolution.

9.5   Risks Associated with Energy Use

The use of energy always accompanies some risk or danger. Even in the use of solar energy, sunlight irradiation can lead to the development of skin cancer from sunburn, and if a dam for hydropower collapses, there will be large numbers of victims. If you ride a car, there is always some risk to have an accident and not a small number of people are killed in a year. In the Great East Japan Earthquake, the energy stored on the earth was released as mechanical energy, and the risks associated with energy generation in nuclear power plants were demonstrated in the worst possible way. The same is true for typhoons and hurricanes. They are not desirable in daily life but necessary for agriculture. They bring water from sea to land, which requires huge energy.

For daily life using energy, people are accepting the risks associated with its use and compensating for it with an insurance system. Regardless of fossil energy or nuclear energy, the risks associated with energy conversion and use are often claimed as if they were zero (safety myths). Without providing reliable information, energy companies have been installing large-scale energy sources in depopulated areas with some economic support in order to cover the large amount of energy consumption in the city area. It is still unclear who will pay how much money due to the accident at the Fukushima nuclear power plant.

Based on the statistics of the death tall per unit of energy output, the risk seems misunderstood with higher risk for larger power use. In reality, the higher the energy density or the power supply, the fewer the number of deaths. This is clear in traffic accidents. i.e., the accident rate decreases with carries using smaller to larger power (energy) as motorbikes, passenger cars, buses, ships to airplanes. Of course, it is well known that equipment using more energy or power, the greater the impact of its accidents, so that more attention is paid to preventing the accident. In recent days most of energy is used to maintain a healthy and cultural life. In order to use energy, it must be accepted that a certain risk is always accompanied in addition to paying the cost directly as money.

It is impossible to sustain people living in large cities only with renewable energy generated in surrounding areas, and hence introduction of huge energy generated in other areas is necessary. This is exactly what today's energy situation is in the city area, and people in big cities do not take the risks associated with energy generation. This should be reminded to urban people when touting the use of renewable energy. Where can wind power generators, mega solar cells, geothermal plants, etc. be installed?

Of course, the risks associated with power generation should be covered by the beneficiaries (energy users). To realize this, energy sources should be installed in city area. This is the right way for long-term energy use. Astro Boy, unfortunately, may not be possible with nuclear energy, but he is showing the ideal form for energy use, not only for him but whole town he is living in, the energy is supported by energy sources in the town. The space station should also stand on this principle. "Installation of a large-scale non-fossil fuel-type energy source in urban area commensurate with their energy consumption" this is the ideal solution for the long-term view. Then a question arises; what is the energy source for that? At present, there is only nuclear energy (nuclear fission and fusion). Building safe nuclear energy plants in the town should be a long-term energy strategy that will last more than a few hundred years. Of course, one can ask, why not build the safe nuclear energy source in a depopulated area. That is right. But the risk cannot be zero. No matter where the energy sources are built, the accompanied risks should be covered by the beneficiaries. It should be avoided to claim unsafety for any energy sources irresponsibly from the city.

Note: Energy production.

The word "energy production" is a very misleading expression. There is a fundamental principle in nature called the energy conservation law. Based on the law, there is no energy production, but energy conversion or transformation of energy form, i.e., conversion of high energy density and power to lower ones beneficial to handle for people. All energy is converted to work and power and finally exhausted as waste heat. Or except for those used as the work, all energy is converted into heat.

In the solar system, hydrogen fusion converts mass into energy (nuclear energy). The energy generated by the fusion is released as EQ and most of them are converted into heat in the sun. So that the sun becomes a star with a surface temperature of about 5800 K. The solar energy is energy emitted from the star with its surface temperature of 5800 K as mostly electromagnetic waves (photons), and reach to the earth, in the order of the wavelength, like X-rays, soft X-rays, ultraviolet rays, visible light, and infrared light. Among them, X-rays and soft X-rays which are hazardous for living beings, are absorbed in the atmosphere, such as ozone generation, and do not reach the ground surface. For this reason, the earth's surface is a paradise for living beings. Photosynthesis is reduction of CO_2 with water using the energy of ultraviolet rays and visible light making carbohydrates, which profit human beings as food energy and bio-energy.

Appendix
Q and A Relating Radiation (EQ: Energetic Quanta)

Radiation is explained in a simple form of Q & A, which also serves as summary.

Q1: What is radiation (EQ: Energetic quanta)?

Q2: What is radioactivity?

Q3: What are sources of radiation (EQ)?

Q4: Are light and radiation (EQ) the same ?

Q5: What are particles that carry energy?

Q6: What kind of particles and light (photons) are included in radiation (EQ)?

Q7: How do energetic quanta (EQ) move in a substance?

Q8: What does radiation exposure (EQ exposure) mean?

Q9: What do the following units related to EQ exposure mean and how they are different with each other? Count rates (cps, cpm, cph), Becquerel (Bq), Gray (Gy) and Siebert (Sv)

Q10: Is the radiation (EQ) exposure of 20 mSv dangerous?

Q11: Does the radiation (EQ) exposure make substances and/or living-beings) radioactive?

Q12: Does a substance glow when exposed to radiation (EQ)?

Q13: What do internal and external exposures mean? What is their difference?

Q14: What happens on radioactive materials ingested into a body?

Question 1: What is radiation (EQ: Energetic quanta)?

Answer 1: "Radiation" originally means physical phenomena of energy emission as light from a material. Above 0 K, any material emits electromagnetic wave (light or photons) with the wavelength of λ corresponding to its temperature (T), i.e. $kT \approx hc/\lambda$, where k, h and c are the gas constant, the Plank's constant and the velocity of light, respectively. This phenomenon is called radiation and the ideal case is referred as the black body radiation. The higher the material temperature, the shorter is the wavelength of the emitted light (see Fig. 1.3 in Chap. 1). A surface thermometer that detects the surface temperature of substances including human beings measures the emitted light in the infrared region form the substances. "Radiation cooling" is the phenomena that the surface temperature of a substance becomes significantly

© Kyushu University Press 2022
T. Tanabe, *Radiation: An Energy Carrier*,
https://doi.org/10.1007/978-981-19-1957-2

lower than the temperature of its surrounding atmosphere owing to radiation from the substance as often observed as freezing of the ground surface in dry winter.

Figure 3.6 in Chap. 3 shows the wavelength distribution of light (photon) emitted from the sun and reached to the earth. The highest intensity of the sun light is given with the wavelengths around 500 nm (nanometer), and its wavelength distribution indicates that the surface temperature of the sun is about 5,250 °C. From the measurement of wavelength distribution emitted from a star hundreds of millions of light-years away, one can determine its surface temperature, internal temperature, and what they are made of. (Reference: https://en.wikipedia.org/wiki/Radio_astronomy)

A radioactive material radiates energy as either or both of particles or photons with their energy roughly above 1 keV equivalent to around 10^7 K. In quantum physics, particles and electromagnetic waves are quantized as quantum particles and photons, respectively. And both in high energy states are unified as the energetic quanta (EQ). **Therefore, throughout this book, the term of "Energetic Quanta (EQ)" is used instead of "radiation".**

EQ can be counted one by one as EQ intensity, i.e. the number of EQ emitted from its source (a radioactive material), is accounted with the SI unit of Becquerel (Bq). A material that emits one energetic quantum per second is called a radioactive material with the radioactivity of 1 Bq or an EQ source having the intensity of 1 Bq (see Sect. 1.3 in Chap. 1).

Question 2: What is radioactivity?

Answer 2: As given in A1, "radiation" originally means the emission of electromagnetic waves from materials at higher temperatures. In fields of radiation physics, chemistry, biology, and medicine, "radioactivity" is used to represent the ability or phenomenon to emit high energy particles and electromagnetic waves referred as energetic quanta (EQ). Materials that emit EQ are called radioactive materials or materials having radioactivity. It should be noted that the "radioactivity" is often used instead of the radiation. In such cases, the radioactivity means the intensity of radiation or EQ. Quantitatively, the intensity of EQ or the radioactivity is compared with using the SI unit called Becquerel (Bq), which indicates how many EQ (energetic particles and photons) are emitted from the EQ source or radioactive material or exposed to a substance (inorganic or organic materials including life-beings) (see Sect. 1.2 in Chap. 1).

Question 3: What are sources of radiation (EQ)?

Answer 3: The radiation (EQ) source means a material that emits energy to surrounding. Electric heater is a radiator or radiation source. As explained in A1, any substance radiates electromagnetic waves (photons), so it is a source of radiation. At the same time any substance absorbs electromagnetic waves. As the balance, only the substance at above room temperature (RT) emits net energy to the surrounding as the radiation. Therefore, in principle, any substance above RT is the radiation source. Depending on the temperature of the substance, the wavelength of the emitted photons changes as shown in Fig. 1.3 in Chap. 1.

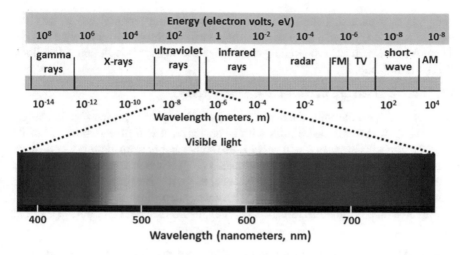

Fig. A.1 Energy and wavelength of electromagnetic waves

The sun is, of course, one the most intensive radiation sources. Stars that emit energy by themselves, including the sun, radiate energy as EQ produced by nuclear reactions in them with very wide energy range and those coming to the earth are recognized as cosmic rays. Usually, in discussion of the effects of radiation, those radiate EQ with energy above a few keV are called the EQ (radiation) source.

Most of elements in the periodic table consist of several isotopes, which are different in their mass number. Most of the isotopes are stable, i.e. they exist permanently as they are. However, some isotopes called as radioisotopes (RI) are not stable and disintegrate radiating EQ to be a stable isotope. In nature, there are numbers of radioisotopes as the radiation (EQ) sources (see Fig. 1.5 in Chap. 1).

Owing to recent technological development, artificial EQ sources are generated. They are accelerators of ions and electrons, X-ray generators, and nuclear reactors, in which EQ are used for various purposes under strict safety regulations (see Sects. 3.3 and 3.5 in Chap. 3).

Question 4: Are light and radiation (EQ) the same?

Answer 4: Yes, both are the same. Radiation (EQ) are either electromagnetic wave (light) or particles carrying energy. Energy, ε, carried by the electromagnetic wave is inversely proportional to its wavelength, λ, or proportional to the frequency, ν, i.e. $\varepsilon = hc/\lambda = h\nu$, where c is the speed of light and h the Planck's constant. In Table 1.1 in Chapt. 1 and Fig. A.1, electromagnetic waves are arranged in the order of the energy/wavelength. Depending on the wavelength they are named differently and used for different purposes. From the longer wavelength they are called as meter (m) wave, millimeter (mm) wave, and micro (μm) wave, infrared light in the wavelength range of about one micrometer, visible light from 0.7 to 200 nm, invisible ultraviolet light above 200 nm. The electromagnetic waves with wavelengths shorter than nm are usually called EQ (radiation). Still they are distinguished soft and hard X-rays, and γ-ray or γ-photon. The electromagnetic waves having very high energy acts like a particle, and very high energy particles behave like light, so they are called quanta. Therefore, the term EQ (energetic quanta) is used to represent radiation in this book (Sect. 1.2 in Chap. 2).

Question 5: What are particles that carry energy?

Answer 5: A particle with a mass of m and a velocity of v carries kinetic energy, $\varepsilon = \frac{1}{2}mv^2$. If the particle collides with a substance, it will give some or all of its energy to the substance, or it will get (absorb) some energy from the substance. Since energy caried by energetic particles or quanta (EQ) is much larger than the kinetic energy of constituent atoms of the substance, EQ will give (deposit) some of their energy to the substance at the collision. This energy deposition or absorption process in the substance is called as radiation (EQ) exposure (see Sect. 1.2 in Chap. 1, and A7).

Question 6: What kind of particles and light (photons) are included in radiation (EQ)?

Answer 6: EQ consist of energetic particles like α-and β-particles, and electromagnetic wave like γ-photons. Figure 1.8 in Chap. 1 shows how energy is deposited to a substance exposed to EQ and how deep EQ can penetrate in it. α-particles can penetrate only very short distance in the substance like the thickness of the skin, and all the energy is deposited. β-particles can penetrate a thickness of several μm. Therefore, exposure to α-or β-particles from the outside does not give large effect on human body. Of course, if the intensity of EQ or absorbed energy (dose) is high, it will burn skin. If EQ sources are taken into a human body, they give internal exposure (See Q13) which is quite dangerous. Since γ-photons are highly transparent, it penetrates thorough the body. Of course, it gives energy while penetrating, i.e. the whole body is exposed to. In most cases, exposure to EQ is given by the γ-photons (see Chap. 3).

Question 7: How do energetic quanta (EQ) move in a substance?

Answer 7: Any EQ move straight until they collide with the constituent elements (atoms and electrons) of the substance and give all or part of their energy. After the collision, EQ change their moving direction, i.e. the trajectory of EQ bends at each collision. Usually the larger the energy transfer at the collision, bending of the trajectory is more significant. Different from water and/or air, EQ move forward until the collision, but do not turn around and collide with matters behind. Compare to the trajectory change of energetic particles, the trajectory change of γ-photons is quite small. Therefore the γ-photons run without significant bending for long distance and spread very gradually (see Sect. 2.4 in Chap. 2, Sects. 4.2 and 4.4 in Chap. 4).

Question 8: What does radiation exposure (EQ exposure) mean?

Answer 8: EQ exposure means that a substance is exposed to EQ and accordingly some or all of EQ energy is deposited to or absorbed in the substance. In case the substance is a huma body, at first the energy is deposited to constituent elements (molecules, atoms, and electrons) in cells resulting their damages, which, in turn, could bring diseases in tissues and organs, or canceration. The deposited (or absorbed) energy is finally converted to heat in the substance. The absorbed energy is accounted as absorbed dose with the unit called as Gray (Gy). 1 Gy means that 1 J (1 J = 0.24 cal.) of energy is absorbed in the mass of 1 kg, i.e. 1 Gy = 1 J kg^{-1} (see Chap. 2).

Question 9: What do those units related to radiation (EQ) exposure mean and how they are different with each other? Count rates (cps, cpm, cph), Becquerel (Bq), Gray (Gy) and Siebert (Sv).

Answer 9: Becquerel (Bq) is used to show the intensity of EQ: how many EQ are emitted from the source or inject to a substance. Historically, the intensity was accounted as the number of detected EQ, i.e. counts per unit time, cps (in a second), cpm (a minute), cph (an hour). Afterwards, Bq was introduced as 1 cps. When a radioactive material emits one energetic quantum (EQ) in a second, it is referred as the radioactive material having the intensity or radioactivity of 1 Bq.

Gray (Gy) represents energy absorbed (or deposited) in a substance exposed to EQ and called the absorbed dose. 1 Gy means that 1 J of energy is absorbed in 1 kg of mass.

Siebert (Sv) is called as the absorbed dose equivalent. Depending on the type of EQ the absorbed (deposited) energy in the unit mass of the substance is different. Therefore, the absorbed dose, Gy, is converted to absorbed dose equivalent, Sv, with normalization of the difference of the kind of EQ introducing a radiation weighting factor (W_R) with the exposure of γ-photons as a reference. Hence for the γ-photons exposure, 1 Gy is equivalent to 1 Sv. W_R for other kinds of EQ are tabulated in Table 2.1 in Chap. 2. Usually news-media use the unit of micro-Sieverts (1 μSv = 10^{-6} Sv) for the EQ exposure. Exactly speaking, in most case, it should be the absorbed dose equivalent of 1 μSv in one hour (μSv/h). In radiation regulation, the allowed exposure for ordinary persons is recommended to be below 20 mSv (1 mSv = 10^{-3} Sv) in a year. That means the integrated dose rate with time exposed to EQ. In Chap. 1, intensity and energy of EQ emitted from EQ sources are summarized in Table 1.2, while intensity exposed and energy absorbed (absorbed dose) in a substance in Table 1.3.

If the kind and the intensity (in Bq) of EQ are determined on the EQ exposure, conversion of Bq to Sv is possible using radiation weighting factors given in Table 2.1. Consider the exposure to ^{131}I as an example, its radiation weighting factor is given as 2.2×10^{-8} Sv Bq^{-1}, so the absorbed dose (Sv) can be estimated by multiplying this number to the EQ intensity measured (see Sect. 2.4 in Chap. 2).

These units are uses as SI units. However, several decades ago, different units based on CGS were used. They are **R** (Roentgen) as absorbed dose, **rad** as radiation absorbed dose (**rad**), and **rem** (Roentgen equivalent man) as absorbed dose equivalent. The relations among these units and SI units (Bq, Gy and Sv) are summarized in Table 5.2 in Chap. 5.

Question 10: Is the radiation (EQ) exposure of 20 mSv dangerous?

Answer 10: If this question means "Does the EQ exposure with the absorbed dose equivalent over 20 mSv cause cancer or any other disease?", the answer is "I have no other way but to say that I don't know." If you asked how dangerous this exposure, the answer could not be made without the comparison of the risk given by the EQ exposure and other cause. The probability of getting lung cancer by the exposure with 20 mSv/y is much lower than the probability that daily smokers will get. For the exposure of 20 mSv/y, the probability getting thyroid cancer is likely to be higher than that of people who were not exposed to EQ. However, it is almost impossible to show how high quantitively the probability is. Human beings are exposed to natural radiation with the absorbed dose equivalent of about 2.4 mSv per person per year. The effect of this exposure cannot be evaluated. In addition, human beings have resilience ability, and the ability varies greatly depending on the individual and his lifestyle. Exposure of 20 mSv/y is not a level that affects individuals, but can only be evaluated as the probability that one or two people in thousands will be affected.

In order to see the effects of one or two people in thousands of people, we have to investigate the irradiation effects on more than tens of thousands or millions of people being exposed with about 20 mSv/y. Obviously, it is not possible to conduct such large number of surveys. After World War II, radioactive materials produced by nuclear bomb tests in atmosphere were dispersed all over the world and air dose rate was more than 10 times higher than the current level. Nevertheless, the effect of the exposed dose is not clear. With the increase of life expectancy, the canceration rate has increased significantly, which makes the estimation of the probability of the canceration owing to the air exposure mostly impossible. (It does not mean that there was no impact.) In addition, since the daily life style influences one's health condition, those who emphasize the safety of exposure to 20 mSv would claim that the exposure does not give any influence, while those who emphasize the risk of EQ exposure would stress the absolute number (not the probability) of people getting influenced by the exposure. Of course, there is no doubt about that the exposure should be kept as low as possible. Because it is not possible to erase the exposure history in past, there is no other way but to try to live a healthy life both physically and mentally, believing that your exposure level will not affect your health (see Chaps. 2 and 4).

Question 11: Does the radiation (EQ) exposure make substances and/or living-beings radioactive?

Answer 11: Not at all. Some people misunderstand that substances exposed to EQ will become radioactive or contaminated, but that is completely wrong. Contamination of a substance occurs only when a radioactive material is transferred to, attaches on, or is taken up in it (see Sect. 2.6 in Chap. 2).

Question 12: Does a substance glow when exposed to radiation (EQ)?

Answer 12: The answer is both "Yes" and "No". Fluorescent or phosphorescing materials are used to display passages to emergency exits and emergency exits. They emit visible light when irradiated with ultraviolet or blue light (shorter wavelength than visible light). These materials also emit the light under EQ exposure. When using X-rays in medical applications, a small fluorescent screen is used to detect where the X-rays are.

Any substance above 0 K emits electromagnetic waves, and the wavelength is shorter for high temperature. When energy is given to a substance, its temperature increases. Then the substance emits the electromagnetic wave shorter than that emitted at the original temperature. This is the original meaning of "radiation". If energy given to a substance is large enough to increase its temperature above around 800 K, one can see the substance glowing (looks red colored). Since EQ give energy to a substance, the substance emits excess energy as the electromagnetic wave. If it is in visible wavelength region, the material glows as fluorescence. So, the answer to the question is "Yes" in respect of energy release, but "No" if the emitted electromagnetic wave is invisible. The energy of light emitted as the fluorescence is usually not high enough to affect the human body (Sect. 1.3 in Chap. 1).

Question 13: What do internal and external exposures mean? What is their difference?

Answer 13: The external exposure means that a substance or a human body is exposed to radiation (EQ) of which source is outside, while the internal exposure occurs when the source is inside. The internal exposure is caused by radioactive materials (mostly radioactive isotopes) ingested and stored in tissues or organs in a human body. For example, internal exposure caused by Iodine-131 (^{131}I) taken into the thyroid gland gives high risk to the development of thyroid cancer, which is most concerned to give radiation effects, especially in children (sect. 2.5 in Chap. 2).

For the external exposure, α- and β-particles are hardly reach organs in a huma body. For the internal exposure, on the other hand, the organs are directly exposed to them. Therefore, the internal exposure by radioisotopes emitting α- and/or β-particles is much hazardous than the external exposure by them. See also A14.

Question 14: What happens on radioactive materials ingested into a body?

Answer 14: Since any radioactive material consists of or includes radioisotopes, its chemical behavior in the body can generally be investigated with using their stable isotopes. Since the chemical behavior of a radioisotope is nearly the same as its stable isotopes, except isotope effects, i.e. some difference mainly caused by their mass difference, any radioisotope ingested in a human body is excreted by metabolism with the same way as its stable isotope. The excretion occurs exponentially with a specific half-life called the biological half-life. Table 5.1 in Chap. 5 summarizes the biological half-life of the main radioisotopes. ^{131}I tends to accumulate in the thyroid gland and is feared to cause thyroid cancer, but it will be excreted about 1/8 in one year after the ingestion. The biological half-life of Strontium-90 (^{90}Sr), which has been widely distributed in atmosphere due to the nuclear tests in the 50s to 70s, is rather long with 49.3 years. That is because the chemical behavior of Sr resembles that of Calcium (Ca), one of the dominant constituent elements of bones and accordingly once it is incorporated into the bones, it is not easily eliminated. In order to enhance the ejection of a radioactive isotope ingested into the body, the most effective way is to replace it by ingestion of its stable isotopes. For example, in the case of ^{131}I taken up into the thyroid gland, its stable isotope of ^{127}I can be ingested as a tablet of sodium iodide (NaI) and/or potassium iodide (KI). If exposure to radioactive iodine is expected like the occasion of nuclear accident, it is recommended to take the iodine tablet (see Chap. 5).

Bibliography

Since the contents of this book are basically well-known facts described in textbooks and technical books, only references necessary to reprint figures are specified in each chapter. However, since there are many good books from which the author has got useful information and knowledges, some of them are given in the following. Figures and tables got permission for reprinting, not requiring the permission, or open for public in web are indicated in their captions showing their origin. The others are the author's own.

As for radiation, so many books, introductory book, textbooks, reference books, technical books, etc. have been published. Standing positions of their authors are various, such as, opposing or promoting nuclear power, standing on pure science, etc. The following are lists of published books, with separation of 7 categories; (a) introductory, (b) radiation and radioactivity, (c) radiation biology, (d) radiation physics and chemistry, (e) radiation measurement, (f) radiation hormesis, and (g) utilization of radiation. All seem to be written based on scientifically correct points of view. Nevertheless, they are not necessarily recommended by the author. And there are lots of good books published but not refereed here. On radiology, radiation oncology and radiation therapy, many more books have been published. However, they are out of scope of this book and not referred here.

Recent days, there appear many web sites managed by governments or public organizations, for examples,

http://www.env.go.jp/chemi/rhm/h27kisisoshiryo.html
https://www.epa.gov/radiation
https://medlineplus.gov/radiationexposure.html
https://www.cdc.gov/nceh/radiation/health.html
https://www.nrc.gov/reading-rm/basic-ref/students/for-educators/09.pdf
https://www.env.go.jp/en/chemi/rhm/basic-info/index.html
https://www.env.go.jp/en/chemi/rhm/basic-info/1st/index.html

Many books written in English are also found easily in any website search systems. Most of them are on radiation measurements, radiation biology and radiology. The latter two focus to risks and effects caused by the exposure of radiation (referred as energetic quanta (EQ)) based on absorbed dose or dose equivalent and medical treatments.

Following are some English books which handle similar subjects with this book.

R.P. Gale, E. Lax, *Radiation: What It Is, What You Need to Know, Vintage*; Illustrated edition (October 8, 2013). ISBN-10: 0307950204, ISBN-13: 978-0307950208

M. Miladjenovic, *Radioisotope and Radiation Physics: An Introduction* (Elsevier Science, December 2, 2012), pp. 254, e-book. ISBN 0323158919

I. Obodovskiy, *Radiation: Fundamentals, Applications, Risks, and Safety* (Elsevier Science, March 9, 2019), pp. 720, e-book, ISBN 0444639861

R.A. Johnson, *Physics of Radiation Effects in Crystals* (Elsevier Science), pp. 736, e-book, ISBN 0444598227

A. Charlesby, *Atomic Radiation and Polymers: International Series of Monographs on Radiation Effects in Materials* (Elsevier Science, June 6, 2016), e-book, ISBN 1483181308

S. Yip, *Nuclear Radiation Interactions* (World Scientific Publishing Company, October 24, 2014), e-book, ISBN 9814644579

In the original Japanese edition, following books (mostly written in Japanese) are sited. Because Japan is the only nation suffered by nuclear bomb, many Japanese books related radiation have been published. For those who would be interested in, they are sited here with their ISBN number for easy searching.

(a) Introductory

H. Iida, K. Anzai, *Etoki Houshasen no Yasashii Chisiki (Easy understanding of radiation with figures)* (Ohm, Ltd, January 1984). ISBN-10: 4274020908, ISBN-13: 97 8-4274020902 (Japanese)

S. Kndo, *Hito wa Houshasenn ni Naze Yowaika, Sukoshi no Houshasen wa Shinpai Muyo (Why people are sensitive to radiation, no need to worry about weak radiation)*, 3rd edition (Blue Bucks) Kindle edition (Kodansha, December 1998), ASIN: B00OB7MDVM (Japanese)

Y. Tateno, *Houshasen to Kenko, (Radiation and Health)* (Iwanami, August 2001). ISBN-10: 4004307457 ISBN-13: 978-400430 7457 (Japanese)

K. Saito, *Shitte Okitai Houshanou no Kiso Chishiki (Basic Knowledge of Radioactivity)* (SB Creative, May 2011). ISBN-10: 4797365684, ISBN-13: 978-4797365689 (Japanese)

J. Tada, *Houshasen; Houshanou ga Yoku Wakaru Hon, (A book for easy understanding of radiation and radioactivity)* (Ohm Ltd., July 2011). ISBN-10: 4274210626, ISBN-13: 978-42 74210624 (Japanese)

Japan Isotope Association Ed., *Houshasen no ABC, (ABC of radiation)*, (April 2011). ISBN-10: 4890732128, ISBN-13: 978-4890732128 (Japanese)

K. Noguchi (ed.), *Houshasen ga Yoku Wakaru Hon, (A book for easy understanding of radiation)* (Popula, Ltd., April 2012). ISBN-10: 459112830X, ISBN-13: 978-4591128305 (Japanese)

T. Araki, *Hibaku no Tadashii Rikai, Houshasen to Houshanou to Houshasei Busshitu towa Dou Chigaunoka? (Correct understanding of radiation exposure—How different are radiation, radioactivity, and radioactive materials?)* (Inner Vision, Ltd., December 2012). ISBN-10: 4902131242, ISBN-13: 978-4902131246 (Japanese)

H. Natori, *Houshasen wa Naze Wakarinikuinoka - Houshaen no Kenkou eno Eikyo, Wakatteiru-Koto, Wakakarani-Koto, (Why radiation is difficult to understand—the effects of radiation on health, what you know, what you don't know)* (Apple Publishing, January 2014). ISBN-10: 487177321 ISBN-13: 978-4871773225 (Japanese)

J. Ando, *Houshasen no Sekai e Youkoso, Fukushima Daiiti Genpatsu Jiko mo Fukumete, (Welcome to the world of radiation- Including the Fukushima Daiichi Nuclear Power Plant accident)* (January 2014) ISBN-10: 4860450884, ISBN-13: 978-4860450885 (Japanese)

M. Kikuchi, K. Komine (ed.) and M . Okazaki, (Illustration), *Ichi kara Kikitai Houshasen no Hontou : Ima Shitte Okitai 22 Episodes, (The truth of radiation, 22 episodes to be notified)* (Chikuma Shobo, March 2014). ISBN-10: 4480860797 ISBN-13: 978-4 480860798 (Japanese)

E. Ochiai, *Houshanou to Jintai, Saibou Bunshi Reberu kara Mita Houshasen Hibaku, (Radiation and human body: radiation exposure from the standpoint of cells and molecules)*, Kodansha, Kindle Edition (March 2014). ASIN: B00JDCI6WA (Japanese)

The Japanese Radiation Research Society, Ed. *Hontouno Tokoro Oshiete! Houshasen no Risuku -Hushasen Eikyou Kennkyuusha Kara no Message*, (Tell me the truth! Risk of radiation - messages from researchers on impact of radiation-), Iryo Kagakusha, February 2015, ISBN-10: 4860034546, ISBN-13: 978-4860034542. (Japanese)

(b) Radiation and Radioactivity

M. Saito, *Houshanou wa Kowainoka—Houshasen Seibutsugaku no Kiso, (Is "radioactivity" scary? - Basics of radio-biology)* (Bungei Shunjyu, June 2001). ISBN-10: 4166601776, ISBN-13: 978-4166601776 (Japanese)

Y. Morimoto, T. Hosoda, J. Hosotani, *Houshasen Sokutei to Suuchi no Hontou no Hanashi, (True story of radiation measurements and figures)* (Takarashimasha, October 2011). ISBN-10: 4796686606, ISBN-13: 978-4796686600 (Japanese)

Y. Yakubukuro, F. Yatagai, *Ima Shiritai Houshasen to Houshanou, Jintai eno Eikyo to Kankyou deno Furumai (Radiation and radioactivity you want to know now -Effects on the human body and behavior in the environment-)* (Ohm, Ltd., December 2011). ISBN-10:4274211444, ISBN-13:978-4274211447 (Japanese)

J. A. Eddy, *The Sun, The Earth, and Near-earth Space, Guide to the Sun-Earth System* (Books Express Publishing, December 2009). ISBN-10: 1782662960, ISBN-13: 978-1782662969 (Japanese)

R.P. Gale, E. Lux, M. Asanaga, *Houshasen to Reisei ni Mukiaitai Minasan e - Sekaiteki Ken-i no Tokubetsu Kougi, (Radiation: What it is. What you need to know. - Special Lecture on Global Authority)* (Hayakawa Shobo, August 2013). ISBN-10: 4152093935, ISBN-13: 978-4152093936 (Japanese)

(c) Radiation Biology

T. Hishida, *Houshasen Igaku to Seimei no Kigen, (Radiology and origin of life)* (Yuhisha, Ltd., June 2004). ISBN-10: 486030053X, ISBN-13: 978-4860300531. (Japanese)

Japanese Society of Radiological Technology (Supervisor), Y. Ejima, H. Kimura (ed.), *Houshasen Seibutsugaku, (Radiation Biology),* 2nd Ed. (Ohm, November 2011). ISBN-10: 4274211193, ISBN-13: 978-4274211195 (Japanese)

N. Sugiura, H. Yamanishi, *Houshasen Seibutsugaku, (Radiation Biology)* 4th ed. (Iryo-Kagakusha, June 2013). ISBN-10: 4860450841, ISBN-13: 978-4860450847 (Japanese)

N. Kubota, *Shihan, Houshasen Seibutsugaku, (New Edition, Radiation biology)* (Iryo-Kagakusha, December 2015). ISBN-10: 4860034651 ISBN-13: 978-4860034658 (Japanese)

Y. Matsumoto (ed.), *Jintai no Mekanizumu kara Manabu Houshasen Seibutsugaku, (Radiation biology, learning from the mechanisms of a human body)* (Medical View, Inc. February 2017). ISBN-10: 4758317259, ISBN-13: 978-4758317252 (Japanese)

K. Komatsu, *Gendaijin no Tame no Houshasen Seibutsugaku, (Radiation biology for modern people)* (Kyoto University Press, March 2017) ISBN-10: 4814000847, ISBN-13: 978-4814000845 (Japanese)

(d) Radiation physics and chemistry

As shown in the text, the most important thing in understanding the interaction between radiation (EQ) and the human body is not direct collisions between EQ and atoms and molecules, but chemical reactions and biological reactions caused by ions and radicals generated by electron excitation in cells. Therefore, the number of published textbooks on radiation chemistry is much larger that on radiation physics. Books written from the viewpoint of energy conversion like this book is quite seldom. In addition, since the recent advances of lasers technologies allow to give so high power as that given by EQ in localized area, photochemistry using lasers are nothing but radio-chemistry and hence important for understanding of EQ exposure effects.

M. Ebara, *Gendai Houshakagaku, (Modern radio-chemistry)* (Kagakudoujin, December 2005). ISBN-10: 4759810447 ISBN-13: 978-4759810448 (Japanese)

S. Kawamura, Y. Arai, K. Kawai, O. Inoue, *Houshakagaku to Houshasenkagaku, (Radio-chemistry and radiation chemistry),* 3rd ed. (Tsusho-sangyo Kenkyusha, March 2007), ISBN-10: 4860450167, ISBN-13: 978-4860450168 (Japanese)

J. Tada, *Wakariyasui Houshasen Buturigaku, (Easy understanding of radiation Physics)* (Ohm, Ltd., 2nd ed. February 2008). ISBN-10: 4274204944, ISBN-13: 978-4274204944 (Japanese)

H. Torii, K. Shodagawa, Y. Watanabe, K. Nakagawa, *Houshasen o Kagakuteki ni Rikaisuru, Kiso kara Wakaru Toudai Kyouyou no Kougi, (Basic understanding of radiation, Lecture in the University of Tokyo)* (Maruzen Publishing, October 2012). ISBN-10: 4621085972, ISBN-13: 978-4621085974 (Japanese)

N. Otsuka, M. Nishitani, *Q&A Houshasen, (Radiation Q&A),* 2nd ed., (Kyoritsu Shuppan, February 2015). ISBN-10: 4320035925, ISBN-13: 978-4320035928 (Japanese)

Japanese Society of Radiological Technology (Supervisor), S. Higahsi, N. Kubo (ed.), *Houshakagaku, (Radiochemistry),* 3rd ed. (Ohm, November 2015) ISBN-10: 4274218147, ISBN-13: 978-4274218149 (Japanese)

(e) Radiation measurement

Y. Utsunomiya, *Zukai Nyumon; Yoku Wakaru Saishin Senryoukei no Kihon to Tsukurikata, (Graphic introduction on basics of a dosimeter and how to make it)* (Shuwa System, February 2013) ISBN-10: 4798037273, ISBN-13: 978-4798037271 (Japanese)

Japanese Society of Radiology (Supervisor), M. Nishitani, K. Yamada, H. Maekoshi (ed.), *Houshasen Keisokugaku, (Radiation Measurements)*, 2nd ed., (Ohm, November 2013) ISBN-10: 4274214699, ISBN-13: 978-4274214691 (Japanese)

K. Furuno, *Houshasen Sokutei no Kiso, (Basics of Radiation Measurement)* (Soeisha/Sanseido Shoten, March 2017). ISBN-10: 488142100X ISBN-13: 978-4881421000 (Japanese)

A. Noutomi, *Houshasen Keisokugaku, (Radiation measurements)* (Kokusai-Bunkensha, March 2015) ISBN-10: 4902590417, ISBN-13: 978-4902590418 (Japanese)

(f) Radiation hormesis

K. Fujino (ed.), *Daishizen no Shikumi, Houshasen Horumisisu no Hanashi -Karada ga Karada o Naosu Saibounai Jihatu Chiryo no Jidai ga Kita-, (Mechanisms in nature, The story of radiation hormesis: Resilience of damages given by radiation exposure)* (Seseragi Publishing, April 2004) ISBN-10: 4884161335, ISBN-13: 978-4884161330 (Japanese)

M. Akamatsu, H. Nakamura, *Houshasen Horumisisu Hayawakari ; Minna Shiranai Tei-sennryo Houshasen no Power, (Understanding of radiation Hormesis : The Power of low-dose radiation exposure that everyone does not know)* Kindle Edition (Amazon Services International, Inc., April 2017) ASIN: B06ZYWLPDT (Japanese)

S. Shizuyo, *Fukushima eno Messeji, Houshasen o Osorenaide!, (Message to Fukushima, Don't be afraid of radiation!)* (Gentosha, February 2017) ISBN-10: 434491113X, ISBN-13: 978-4344911130 (Japanese)

(g) Utilization of radiation

A lot of books have been published for medical uses of radiation, only a few are given here.

T. Iida, *Senshin Houshasen Riyou, (Advanced use of radiation)*, Osaka University Press, March 2005, ISBN-10: 4872591895, ISBN-13: 978-4872591897 (Japanese)

Japanese Society of Radiochemistry (ed.), *Houshasenn-kagaku no Susume – Densi, Ion, Hikari no Biimu ga Kurashi wo Kaeru, Sangyo wo Kaeru, (Association of radiochemistry: The beams of electrons, ions, and light change lives and creating industries)* Academic Publishing Center, April 2006, ISBN-10: 4762230502, ISBN-13: 978-4762230509. (Japanese)

K. Higashijima, *Houshasen Riyou no Kiso Chisiki, Handoutai, Kyouka Taiya kara Hinshu Kairyo, Shokuhin Shosha made, (Basic knowledge of radiation to use for semiconductors, reinforced tires, breeding, and food irradiation)* Kodansha, October 2006, ISBN-10: 406257518 3, ISBN-13: 978-4062575188. (Japanese)

H. Kudo, *Gennsiryoku Kyoukasho, Houshasen Riyo, (Textbook of Nuclear Engineering, Radiation Usage)* (Ohm, February 2011). ISBN-10: 4274209849, ISBN-13: 978-4274209840 (Japanese)

Y. Kubo, *Yakugaku ni Okeru Houshasen Houshasei Bushitsu no Riyo, (Use of radiation and radioactive materials in pharmacy)*, 3rd ed., (Kyoto Hirokawa Shoten, February 2012). ISBN-10: 4901789821, ISBN-13: 978-4901789820 (Japanese)

Printed in the United States
by Baker & Taylor Publisher Services